Foundations of Engineering Mechanics

The series "Foundations of Engineering Mechanics" includes scientific monographs and graduate-level textbooks on relevant and modern topics of application-oriented mechanics. In particular, the aim of the series is to present selected works of Russian and Eastern European scientists, so far not published in Western countries, drawing from the large pool of experience from major technological research projects. By contributing to the long tradition of enrichment of Western science and teaching by Eastern sources, the volumes of the series address to scientists, institutional and industrial researchers, lecturers and graduate students.

Valentyn Skalskyi · Zinoviy Nazarchuk · Olena Stankevych

Acoustic Emission

Fracture Detection in Structural Materials

Springer

Valentyn Skalskyi
Karpenko Physico-Mechanical Institute
of the National Academy of Sciences
of Ukraine
Lviv, Ukraine

Zinoviy Nazarchuk
Karpenko Physico-Mechanical Institute
of the National Academy of Sciences
of Ukraine
Lviv, Ukraine

Olena Stankevych
Karpenko Physico-Mechanical Institute
of the National Academy of Sciences
of Ukraine
Lviv, Ukraine

ISSN 1612-1384 ISSN 1860-6237 (electronic)
Foundations of Engineering Mechanics
ISBN 978-3-031-11290-4 ISBN 978-3-031-11291-1 (eBook)
https://doi.org/10.1007/978-3-031-11291-1

This Springer imprint is published by the registered company Springer Nature Switzerland AG
The registered company address is: Gewerbestrasse 11, 6330 Cham, Switzerland

Preface

The problem of technogenic and environmental safety of significant objects of industrial and social infrastructure is becoming more and more urgent both for many economically developed countries and for the world in general.

During long-term operation under the influence of various physical factors, structural materials degrade, i.e., change their basic physicochemical and physicomechanical characteristics. The consequences of this phenomenon are especially dangerous for the objects of the energy complex, chemical industry, pipeline transport, machine building, and bridge structures, where there are factors of aggressive working environment (corrosion, hydrogen, and radiation embrittlement, etc.). Degradation of structural materials contributes to the initiation and development of the most dangerous defects such as cracks, which can cause spontaneous fracture of structures, which, in most cases, lead to significant material damage, environmental, and human casualties losses.

The most important problem of technical diagnostics (TD) and nondestructive testing (NDT) for the objects of long-term operation is to determine their residual life and establish such period and conditions under which they can be safely operated. According to the concept of "safe damage", the presence of cracks may not always be the reason for removing the product or some structural element from the service. To ensure their normal functioning, it is necessary to be able to identify the critical parameters of the crack and the mechanisms of its propagation, which characterize main points of fracture: the moment of start, subcritical growth, and transition of fracture to the supercritical stage.

It is possible to determine the level of wear of structural materials and choose methods for determining the residual life of relevant objects and products by determining the types of their fracture. Such data can be obtained in the laboratory by destructive test methods; however, they are non-effective for TD of the elements of structures and products in the operating conditions.

At present, the method of acoustic emission (AE) is effective for studying the dynamics and stages of defects growth in the structure of materials and assessing the real state of objects during their TD. The possibility to use it to assess the damage of materials is based primarily on correct physical models of elastic wave radiation.

They provide solving the corresponding non-stationary dynamic problems of crack theory in terms of displacements. Their solutions make it possible to determine the expected amplitudes of the AE electric signals generated during initiation or propagation of fracture. At the same time, it serves as a basis for the creation of original methodological approaches and means of selection and processing of AE signals with the appropriate software.

Along with theoretical research, the development of methodological bases of AE-diagnostics with the use of new approaches to the processing of AE signals to identify the types of fracture and their experimental confirmation is an important problem. The Fourier transforms are mainly used to obtain information about the spectral components of the signal. At the same time, from the standpoint of analysis of any signals and functions in the frequency domain and their exact recovery after transformations, the Fourier series expansion of signals has a number of disadvantages and limitations: having good localization by frequency, it has no temporal resolution; limited informativeness of the analysis of non-stationary signals and almost complete lack of possibility to analyze their specific features (singularities); it gives global information about the frequencies of the studied signal and does not give an idea of its local properties with rapid time changes in the spectral composition. If it is necessary to perform time localization of spectral components, one should refer to the frequency–time representation of signals. The wavelet transform (WT) is one of the transformations that provide this representation.

The monograph presents topical theoretical and experimental studies aimed at the creation of modern methods for diagnosing macrofracture of solids and assessing of the structural materials state. New mathematical models of displacement fields due of the crack-like defects development in materials and a new criterion for assessing types of macrofracture of materials by energy parameters of WT of AE signals are built.

Lviv, Ukraine

Valentyn Skalskyi
Zinoviy Nazarchuk
Olena Stankevych

Contents

Abbreviations

δ_{j3}	Kronecker symbol
Δ	Two-dimensional Laplace operator
Δ_3	Three-dimensional Laplace operator
ε	Deformation
φ	Scalar potential
γ	Constant
μ	Poisson's ratio
ν	Opening the crack; crack propagation rate
ρ	Material density
σ	Stress
ω	Cyclic oscillation frequency
ω_j	Wave number
ω_p	Central frequency
ξ	Surface point
$\Psi(\omega)$	Fourier transform of wavelet function
$\psi(t)$	Wavelet function
A	Amplitude
$A_m(t)$	Approximating component of discrete wavelet transform
a_{mk}	Approximating coefficient
b	Time shift
a	Crack radius; scale
C_ψ	Normalization factor
c_1	Longitudinal wave velocity
c_2	Transverse wave velocity
$D_m(t)$	Detailing component of discrete wavelet transform
d	Depth of crack
d_{jk}	Detailing coefficient of discrete wavelet transform
E_{MAE}	Energy parameter of the MAE signal
E_S, $E_{total}(t)$	Energy of signal $f(t)$
E_{WT}	Energy (criteria) parameter of local impulse
$f(t)$	Function (signal)

f	Frequency
$f_{\max}$	Peak frequency corresponding to $WT_{\max}$
G	Shear module
g_l	High-pass filter of discrete wavelet transform
$H(t)$	Heaviside function
h_l	Low-pass filter of discrete wavelet transform
J	Number of signal decomposition levels
$J_0(z)$	Bessel function of zero order
K	Some physical quantity
$\hat{\mathbf{K}}$	Complex amplitude of K
$K_{\mathrm{I}}, K_{\mathrm{I}C}, K_{\mathrm{I}AE}$	Stress intensity factor
k_{ψ}	Normalization factor
m	Mass of the sample
N	Signal length; number of load cycles
$\vec{n}$	External normal to the surface S_V
$Ox_1x_2x_3$	Cartesian coordinate system
P	Load
P_{if}	Initial fracture load
P_{ff}	Final fracture load
$P_{jk}(x)$	Helmholz potential
p	Pressure
$R(\rho)$	Rayleigh function
rE_{WT}	Signal energy distribution by levels of wavelet decomposition
r	Distance from crack epicenter to observation point
S, S_1	Plane crack region
S_0	Body surface
$S^{\pm}$	Opposite surfaces of crack
S^*	Area in which displacement $u_3 = 0$
S_V	Boundary surface of isotropic body
S_{Vu}	Body surface with given displacements
$S_{V\sigma}$	Body surface with given stress
$s(t)$	Signal
$T(t)$	Time load function
$t, \tilde{t}, t_*$	Time
Δt	Pulse radiation duration
$U_i(\mathbf{x}, t)$	Displacement function
$\vec{\mathbf{u}}$	Displacement vector
u_r	Radial displacement
u_3	Vertical displacement
$\hat{u}_j^{\pm}$	Displacement of points of crack surfaces $S^{\pm}$
V	Area of isotropic body
Wi	Wavelet packet
$W(a,b)$	Integral wavelet transform
WT	Wavelet coefficient
$W(a,b)$	Wavelet coefficient of continuous wavelet transform

$WT_a(t)$	Approximating function
$WT_{m,n}$	Wavelet coefficient of discrete wavelet transform
$WT_{\max}$	Maximum magnitude of wavelet coefficients
x_1, x_2, x_3	Coordinates of surface point
AE	Acoustic emission
BIEs	Boundary integral equations
CM	Composite materials
COD	Crack opening displacement
CWT	Continuous wavelet transform
DPWT	Discrete packet wavelet transform
DWT	Discrete wavelet transform
FFT	Fast Fourier transform
FWP	Feed water pipeline
HAZ	Heat affected zone
MAE	Magnetoelastic acoustic emission
MSA	Multiple-scale analysis
NDT	Nondestructive testing
SIF	Stress intensity factor
TD	Technical diagnostics
UHP	Ultra-high-pressure power
WT	Wavelet transform

Chapter 1
Macrofracture of Structural Materials and Methods of Determining Its Type

Structural materials from which the components and structures are manufactured must withstand a wide range of loads during their operation. Material ductility is an important characteristic of its reliability, i.e., the ability to absorb mechanical energy due to plastic deformation without fracture. There are many factors that affect the tendency of material to resist fracture. These include high or low temperature, plastic deformation, loading rate, radioactive irradiation, external environment and stress state, scale factor, stress concentrators (e.g., cuts, cracks), etc.

Under the influence of these factors during long-term operation, most of the structural materials become brittle, thus reducing the fracture toughness and, consequently, the time of safe operation of industrial facilities or products. Information about the ductility or brittleness of the material used is crucial for the choice of a method of calculating the residual life of the relevant products and structural elements. Therefore, to increase the efficiency of technical diagnostics (TD) and nondestructive testing (NDT) of industrial facilities and products, it is important to establish the type of material fracture.

This chapter presents the analysis of the types of fracture and their features, the state of research on the application of the method of acoustic emission (AE) and magnetoelastic AE (MAE) to detect micro- and macrofracture of structural materials, as well as the spectral analysis of AE signals by their wavelet transform (WT).

1.1 Types of Structural Materials Fracture

The decrease in the strength of materials is explained by the presence of defects in them, and such defects may be the dislocations that cause plastic deformation at relatively low stresses and cracks, because of which brittle or quasi-brittle fracture can occur. The last one means the fracture without significant plastic deformations. The fracture properly occurs in several stages: the initiation of microcracks, the formation of macrocracks, as well as their propagation throughout the body.

V. Skalskyi et al., *Acoustic Emission*, Foundations of Engineering Mechanics,
https://doi.org/10.1007/978-3-031-11291-1_1

There are number of attempts to grade the types of fracture by different classification features [1, 2]. In general, it is divided into brittle and ductile [3–8]. Cyclic loads (vibration, frequent impacts, etc.), under which the initiation of micro- and then macrocracks develops fatigue of materials associated with the accumulation of damage, promote fatigue fracture. Fracture can in many cases be of a mixed mode, that is, it can have the features of both brittle and ductile fracture. Sometimes these features are manifested sequentially, for example, fracture, starting as a ductile, turns into a brittle, or, conversely, a brittle crack during a stop may have the features of a ductile fracture.

For both mentioned above modes of fracture, the mechanism of crack initiation is the same. The cause of microcracks initiation, as a rule, is the accumulation of mobile dislocations (plastic deformation) in front of obstacles (grain boundaries and interphase, possible inclusions) [9].

The main difference between brittle and ductile fracture is the size of the plastic zone near the crack tip. During the brittle fracture in front of the propagating crack, it is insignificant, and during the ductile, it is large and the crack itself is blunted at the tip. Figure 1.1 illustrates the character of tensile curves for the brittle and ductile materials. In both cases, the area under the curve is equal to the absorbed energy for the development of fracture.

The second important difference between the modes of fracture is the crack propagation character [10]. Ductile fracture is caused by the low propagation rate, and during the brittle one, it can run into 10–70% of the speed of sound propagation. Therefore, the brittle fracture is often called “sudden” or “catastrophic”. Under the ductile fracture, the crack faces open, increasing its radius, which occurs under significant plastic deformation and requires a constant expenditure of energy for its propagation (Fig. 1.1). The brittle fracture is more dangerous, because it takes place almost instantly and is difficult to predict.

Fig. 1.1 Tensile diagram under brittle (*1*) and ductile (*2*) fracture

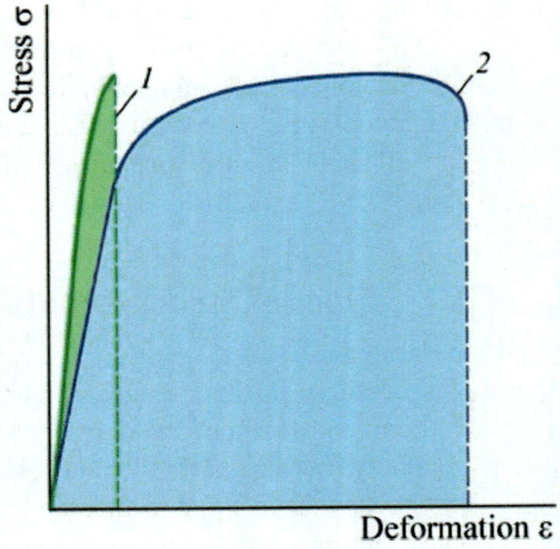

For the brittle materials, the residual deformations under fracture do not exceed 2–5%. The ductile ones are fractured under significant residual deformations, which can reach up to 20–25%. Typical ductile materials include low-carbon steel, aluminum, and brittle materials include cast iron, tool steel, glass, concrete, ceramics, natural stones, etc.

Ductile and brittle materials differ by the character of fracture under tension. The first ones have a greater tear resistance of particles than their displacement relative to each other. They are destroyed mainly by the displacement of particles in the planes of the greatest tangential stresses action. It is due to the displacement of the particles that the length of the sample of ductile material increases under tension, and the fracture site in the neck has the form of a crater, the walls of which are inclined to the sample axis at an angle of 45°. The bottom of the crater is the surface of the initial internal crack, which occurs after the formation of the neck.

Brittle materials, on the other hand, are more resistant to shear than to tear and are destroyed under tension suddenly due to their particles tear along the cross-sectional plane. The phenomena of yield, strengthening, and neck formation on samples manufactured from such materials are absent. The only characteristic feature of brittle materials is the tensile strength.

The division of materials into brittle and ductile is quite conditional, because with the change of the test conditions, the same material can rupture both brittle and ductile. For example, cast iron, which is brittle in tension and twisting, ruptures ductile under compression. Dents on the surface of the brittle marble during the Rockwell hardness tests indicate plastic deformation. For many metals, the character of fracture changes with decreasing temperature [11]. Steel and aluminum at normal temperature, with the correct shape of the parts (rods) under tension, compression, twisting, or bending, have lower resistance to shear than to tear, and the mode of their rupture is ductile. However, at low temperatures, predrawing or in case of their regularly shaped parts, the shear deformation is more complicated. In the more complex stress states, as well as for other reasons, the shear resistance may increase and exceed the tear resistance, because of which the material will fracture brittle. Therefore, it is more correct to speak about the ductile and brittle fracture character of the material. Let us speak in more detail about their features.

1.1.1 Ductile Fracture

Figure 1.2 presents a scheme of possible types of ductile fracture at the macroscopic level. Type *A* fracture is typical of monocrystalline metal samples with a hexagonal lattice, in which sliding along the base plane ends with separation of the sample surfaces by shear (slip) [12]. Monocrystals of some metals with a face-centered cubic fracture with the formation of a neck, which degenerates into a line—"knife blade" (type *B*). In other crystals with the same lattice (e.g., a copper–aluminum alloy), the neck may be practically absent, and the fracture occurs after intense sliding with the appearance of coarse deformation bands (type *C*). In polycrystalline samples

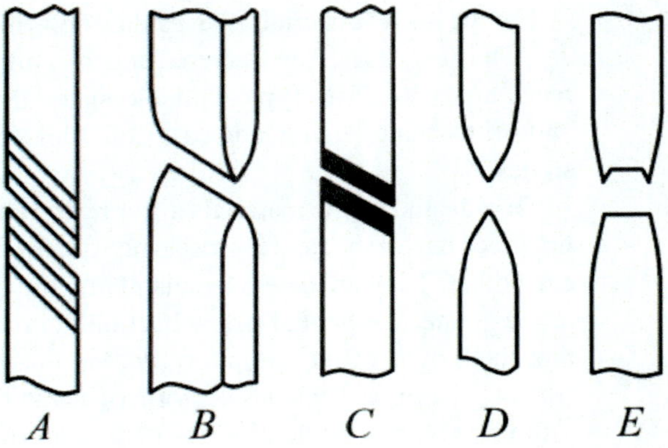

Fig. 1.2 Types of ductile fracture

of high-plastic metals with a face-centered cubic lattice, the narrowing of the cross section before failure reaches 100%, in which case the neck turns into a point (type *D*). In less plastic polycrystalline samples, the most characteristic fracture is formed, and the failure occurs by the "cup-cone" mechanism (type *E*).

The sequence of ductile fracture stages is illustrated in Fig. 1.3. The process begins with the stage of sample narrowing (formation of the neck). Inside the material, small micropores appear which during further deformation grow and coalesce, forming a crack that is perpendicular to the direction of the applied stresses. The crack continues to grow, moving to the lateral outer surfaces of the sample.

The surface along which the boundary crack propagates is located at an angle of 45° to the axis of the sample, where the maximum tangential (shear) stresses act. As a result of the sample shear at the final stage, two fracture surfaces of the "cup-cone" type are formed. Visually, they have a matte fibrous irregular structure that reflects light poorly. Pits are fixed on them, which is a result of microplastic deformation and rupture of walls between micropores (Fig. 1.4a).

Known models of micropores formation at the boundary between inclusions and a ductile matrix of the metal are based both on the principles of continuum mechanics of a solid (the size of the inclusion is more than 1 μm) and on the theory of dislocations. Among the continuum models, the most popular is the Argon's model and its

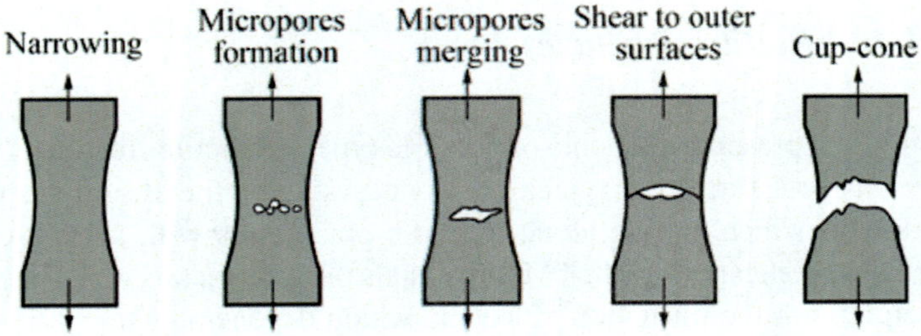

Fig. 1.3 Sequence of ductile fracture stages

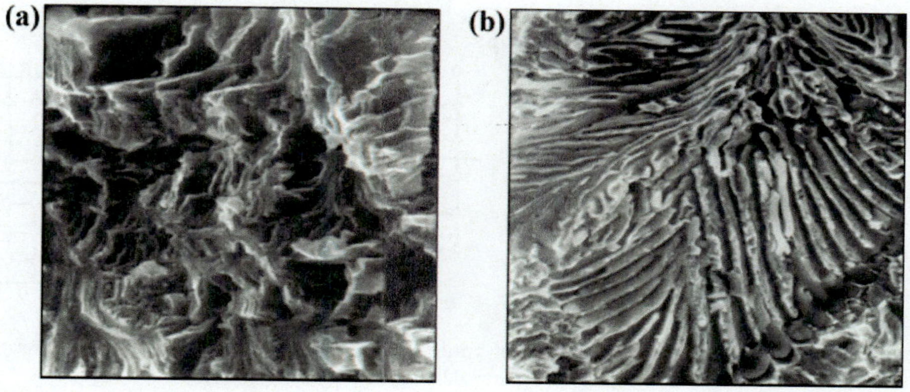

Fig. 1.4 Character of material fracture: **a** dimple relief under ductile fracture and **b** river lines under brittle

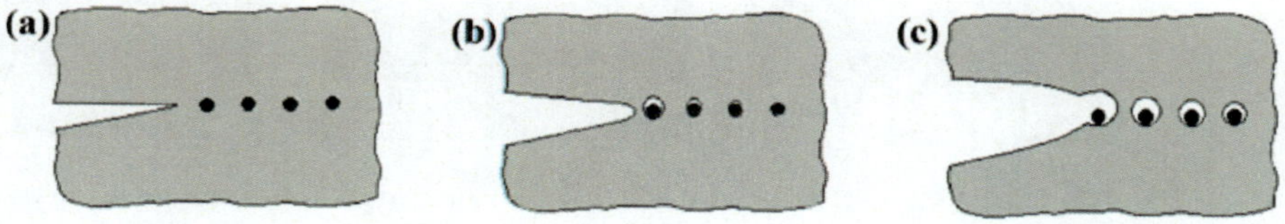

Fig. 1.5 Chart of the micromechanism of a ductile crack growth: **a** inclusions at the crack tip; **b** micropores growth in front of the crack; **c** their coalescence with the tip

modifications [13], while dislocation models of micropores formation at the inclusions are effective in the presence of submicron particles (particle size less than 1 μm) [7]. These models are successfully used to analyze the micromechanisms of ductile fracture of a solid in front of the crack tip by metallography (Fig. 1.5).

1.1.2 Brittle Fracture

At the macroscopic level, during this fracture, the crack propagates almost perpendicular to the direction of the applied stress (Fig. 1.6). After fracture, the fracture surface is relatively smooth, but may, however, have a certain pattern, which depends on the material properties. Very hard materials do not have a pattern, and amorphous types of ceramics and glass have a shiny and smooth surface. Some brittle materials contain "striations and ridges" on the surface, which disperse from the crack origin and propagate in the material (Fig. 1.4b).

Prior to fracture, the material under tension undergoes elastic deformation, sometimes—an insignificant plastic deformation (Fig. 1.6b). When normal stresses reach the value of the tear stresses, the interatomic bonds along a certain plane (the cleavage plane) are alternately broken, and two new interface surfaces are formed (Fig. 1.6c).

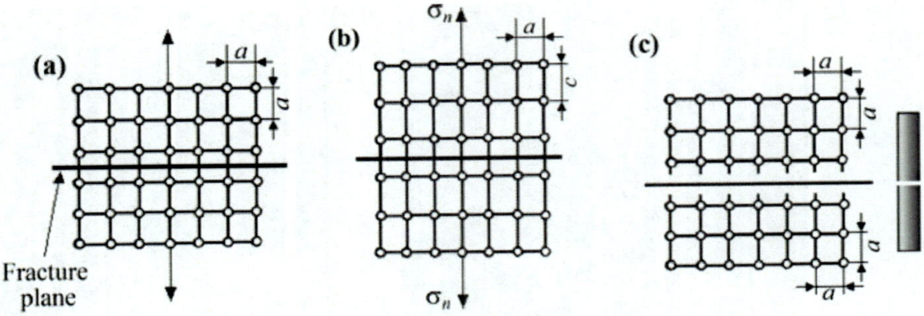

Fig. 1.6 Elementary mechanism of brittle fracture under the action of normal stresses: **a** initial crystal lattice of the metal; **b** its elastic deformation; **c** brittle tear (fracture)

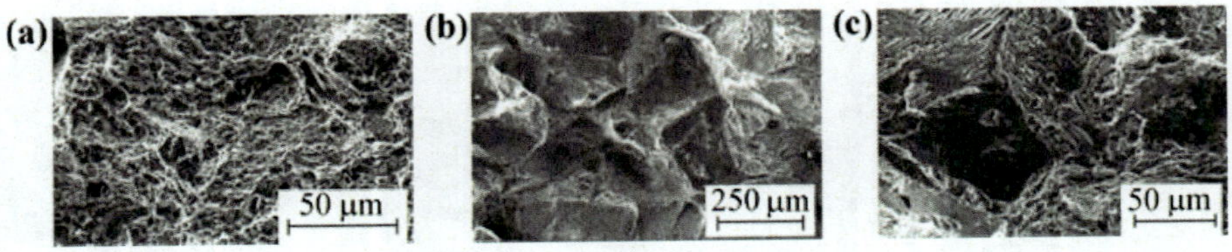

Fig. 1.7 Morphology of the fracture surface of three types of steel: **a** transcrystalline; **b** intergranular, and **c** mixed-mode

The stress required for brittle fracture depends on the bond energy between the atoms. According to the literature [14], for brittle materials it is approximately $E/10$, where E is the modulus of elasticity. However, the experimentally obtained stresses are most often in the range $E/100$–$E/10{,}000$. This low tensile strength is explained by the effect of the stress concentration on microcracks, voids, surface scratches, sharp angles, and so on. The stress amplification factor depends on the shape, size, and orientation of the microcracks. At the microscopic level, the crack under brittle fracture propagates either along the grain body (transcrystalline cleavage) (Fig. 1.7a), or along the grain boundaries (intergranular) (Fig. 1.7b).

1.1.3 Mixed-Mode (Ductile–Brittle or Brittle–Ductile) Fracture

The division of fracture into ductile and brittle is very conditional, as it does not allow us to cover all its characteristics at the micro- and macrolevels. It should be noted that it is usually of a mixed mode (Fig. 1.7c): ductile–brittle or brittle–ductile fracture. For example, there are quite enough plastic materials in which a significant total deformation accumulates at the stage of a crack initiation. Crack growth in this case is characterized by high-speed and low-energy consumption with the formation of a direct fracture surface. Such fracture for energy reasons should be attributed to the ductile mode, and for other reasons it is brittle, although it occurs at a stress higher than the yield strength. From an engineering point of view, such a material will be ductile, and in the case, it already has a crack of sufficient length and it will be very brittle. Under certain conditions, micromechanical character of fracture is possible, which is accompanied by intense plastic deformation, but localized in a small volume of the material. It is characterized by low-energy consumption and for engineering reasons should be classified as brittle, although it leads to the formation of the oblique fracture.

Thus, the terms "brittle" and "ductile" fracture always need to be specified with the indication of some conditional "tolerances" either by the value of the fracture stress relative to the macroscopic yield strength, or by the fracture mode (on a micro- or macroscale), or by the value of absorbed energy or residual deformation, etc.

Most metals and alloys (except for metals with a face-centered lattice) can rupture both ductile and brittle depending on the structural state and the combination of external conditions, among which the most important are the temperature and speed of loading, stress, properties of the loading system and environment.

Regularities of different fracture types of structural materials are studied mainly under tension of smooth (without cracks or other stress concentrators) samples, using tensile curves, the results of metallographic and fractographic studies, by determining crack growth resistance of the material. Unfortunately, such methods are not suitable to identify the types of materials fracture during TD and NDT of the object state in its operation conditions.

1.2 Application of the Acoustic Emission Method to Detect the Fracture of Structural Materials

1.2.1 General Information About Acoustic Emission

The acoustic emission (AE) method, which is based on the registration and processing of elastic waves arising from the formation, change, and fracture of the structure of different materials, is currently the most effective for studying the development of defects in materials and assessing the state of engineering objects during their TD [15–25].

It is known that for metals, the source of AE, in addition to the formation of cracks, can be a few physical phenomena, in particular, the movement of dislocations, the separation of dislocation loops from attachment points, shifts, lattice twinning, phase transitions, voids formation near cracks under plastic deformation, etc., which in certain cases can occur in a stressed structure simultaneously.

Methods of signal processing and determination of their informative parameters depend on the AE type. Discrete and continuous AE are considered. Note that this division is conditional, since the possibility of separating the registration of AE pulses depends on the characteristics of the equipment. In practice, the emission of both types is usually observed. It is known that the subcritical crack growth in metals under the action of external and internal factors is jump like. Long periods of stable state of the crack, when plastic deformation can grow near its tip, alternate with the points of time when the crack changes its length with a speed close to the speed of sound, passing into a new equilibrium state. This transition is associated with a change in the stress state (unloading) of the material in the vicinity of the crack and is accompanied by the emission of an elastic wave, which a transducer registers as a signal of a discrete AE. In the intervals between jumps during the accumulation of plastic deformation, a continuous AE characteristic of plastic deformation is observed. Fatigue cracks propagate similarly. The creep of the material is accompanied by both continuous and discrete AE due to the initiation and propagation of microcracks. During stress corrosion, corrosion cracking is also accompanied by intense flashes of discrete AE. Although in all these cases the crack may not yet be dangerous for the structure, AE indicates the development of a defect, i.e., warns of the danger of catastrophic fracture.

Discrete AE is observed during the technological processes control, where there is a high probability of crack initiation (welding, quenching, diffusion saturation, for example, hydrogenation, etc.), as well as during the study and control of stress corrosion cracking, strength, heat resistance, fatigue failure, friction, and wear. Continuous AE is associated with plastic deformation, corrosion of materials, and other physical phenomena. Since during all these processes, as a rule, both discrete and continuous components of AE can occur, and it is important to be able to separate and identify the corresponding AE signals according to certain special features in order to study the dynamics of fracture.

The informative parameters of certain signals of a discrete AE, signal flows, and parameters of a continuous AE are distinguished. In particular, AE signals for one act of generation by a source are characterized by amplitude, duration, rise time, total count, frequency, and so on. The signals flow can be further characterized by the average frequency of events, amplitude, time and amplitude–time distributions, spectral density, correlation function, average value, and dispersion. Each characteristic is related to a specific physical process that generates the AE and contains information about its development or the state of the object of control.

1.2.2 Some Approaches to the Identification of Fracture Types by the AE Method

According to the results of the literature sources analysis, the existing approaches to the identification of fracture types of materials can be divided into the following groups: analytical or analytical–numerical and experimental.

Analytical (analytical–numerical) methods are based on the development of mathematical models with appropriate differential equations and boundary conditions to describe the initiation and propagation of defects as the AE sources [18, 24]. The established dependences between the parameters of the displacements fields that occur in the body due to the elastic AE waves propagation generated by different sources and the parameters of the sources themselves make it possible to effectively detect both the defect location and its type and dimensions. By identify the type of material fracture in a timely and correct manner, one can prevent catastrophic failure of the object with all its consequences.

In modeling the initiation and growth of cracks, the defect is considered as a fracture surface and for modeling the growth of cavities around different types of inclusions, voids in the ductile material, etc.—as a bulk source. Since a crack is considered to be the most dangerous defect, especially under brittle fracture, more attention is paid to the study of its initiation and propagation. Although the papers study the wave motion due to dislocation processes, changes in pressure in the spherical cavity in material, oscillations of inclusions and scattering of waves on them, etc., are also known [26–35].

Many publications have been devoted to the study of the emission of elastic waves by cracks of different types and configurations located in an infinite space or a half-space [36–61]. For this purpose, the methods of integral transformations of Wiener–Hopf, Kanyar, Fourier, Laplace and Hankel transforms, the finite difference method, and the Green's function are used.

According to the literature, in most dynamic problems about bodies with stationary or growing cracks of different types, the authors mainly determine displacements [26–39, 41–51, 57, 59–62] or stress and displacements [52, 55, 58, 63–70] in the vicinity of the tip, in the center, on the defect surfaces or in its near or far zones. The peculiarity of the approaches used is that it is assumed that the defect is initiated in an infinite body, since the size of the newly formed defect is much smaller than the linear dimensions of the studied body, i.e., displacements on the body surface are not considered.

It is known that when a crack is in an engineering structure, the AE changes significantly due to the interaction of elastic waves with the body boundaries. Then the wave picture is much more complicated than in an infinite body because the interaction of elastic waves with the surface generates surface waves, which significantly affect the displacements redistribution. Considering the external free surface of the body during the mathematical modeling of the wave propagation in it, it is possible to describe physical phenomena more correctly.

For example, in [71], the displacement of the surface due to the fracture in the vicinity of a penny-shaped semi-infinite crack tip under action of tensile loading is studied. The crack growth at a certain angle to the vertical is studied first in an infinite elastic solid, then in a half-space. The parameters of signals that carry information about the crack size, its orientation, and the fracture at the tip are investigated. It is found that the oscillations of the surface in the localized region directly above the crack tip are caused by a longitudinal wave that dominates the transverse one. The Rayleigh wave dominates in the far zone, although the time of its arrival is longer.

Vertical displacements on the free surface of the half-space are determined in [72]. Two cracks are considered: the plane and Y-shaped ones. It is found that in both cases the surface displacements have clearly expressed maxima and minima, which depend significantly on the depth of the crack location and its orientation. Signals from a straight branched crack and a circular cavity are different, which can be useful during NDT.

The method of boundary integral equations (BIEs) proved to be highly effective in solving three-dimensional dynamic problems for bodies with cracks and inclusions [73–76]. In [77, 78], using the BIEs method, the characteristics of the elastic field displacements on the half-space surface initiated by the displacement of opposite surfaces of different type internal cracks are investigated.

Recently, computer modeling of physical processes has become popular, which simplifies the study of their dynamics and features essentially to compare to analytical methods. In particular, the results of the investigation of the defects detection in the elements of the railway track on the simulated AE sources are given in [79]. The AE sources located on the cross section of the railway rail are computer modeled, changing their frequencies and depth in vertical direction. The WT is used to distinguish the frequency–time features and modes of the AE signals. Then, the influence of different distances of AE signals propagation and their features for different types of sources and location depth are studied.

In [80], experimental, analytical, and computational methods are combined to form a finite element model of controlled data for AE generation in the ABAQUS system. The AE is considered as part of the dynamic process of energy release caused by the crack initiation. For this purpose, the experimental data obtained based on digital image correlation are used to construct dependences under tension and to determine the parameters of damage initiation. These dependences are used for 3D finite element modeling. To create a realistic computational model of the AE source, the transition between static and dynamic responses is evaluated. Numerical simulation of AE signals from the dynamic response due to crack growth is analyzed in the context of the inverse problem of identifying the source and determining the influence of the material and geometry of the object on waves propagation.

Experimental methods. To identify the AE sources during experiment, and hence the types of fracture, it is necessary, first of all, to separate AE caused by the microcracks formation and displacement of dislocations. It is methodically difficult to separate AE signals from dislocations and cracks at one object, as the frequency and energy spectra of these signals are unknown. This problem can be simplified to study the parameters of AE from material in which one (main) source of radiation is known in advance. Today, there are several methods for identifying different types of failure, but reliable criteria for the separation of AE caused by one reason or another, unfortunately, are missing.

Known methods of AE-identification of the fracture types are based on the analysis of the AE signal parameters: amplitude and amplitude distribution, time of increase of the pulse leading edge, frequency spectrum [81–88], etc.

In [81], the following types of fracture are identified by the amplitude–frequency characteristics of AE signals: plastic deformation, which is accompanied by a continuous AE with a natural oscillation frequency of approximately 20 kHz; material yield by a discrete AE with an oscillation frequency of about 50 kHz; brittle fracture—in the range up to 250 kHz.

A crack growth model, which establishes a relationship between the extension area of macro- and microcracks and the total count and amplitude of AE signals, is proposed in [82].

The AE under plastic deformation of metals is studied in [83], where it is proposed to describe AE signals from a single deformation act in the form of a vanishing radio pulse.

The influence of the grain size on the AE emitted during plastic deformation of copper is investigated in [84] and compared with the previously obtained results for aluminum. The largest number of events and the energy of AE signals are observed under plastic deformation of samples of copper with a grain size of 70 μm and aluminum with a grain size of 55 μm. AE activity in these materials is caused by cross-slip of dislocations; according to the results of experiments, it is higher in copper than in aluminum.

According to the literature [85], the range of changes of amplitudes is three orders of magnitude, and their distribution can be used to analyze deformation, considering that the initiation and propagation of macrocracks are accompanied by AE signals of much higher amplitude than during deformation.

In [86], the amplitude distribution of AE signals is used for the first time to separate the sources of their radiation under brittle fracture and plastic deformation.

A bimodal distribution of amplitudes, in which the first maximum (at low amplitudes) corresponds to the sliding mechanism and the second one (at three time larger amplitudes) to twinning, is established in [87].

The amplitude distribution of AE signals is presented mainly in the form of a histogram of the AE envelope signals maxima. Such a representation makes sense for a discrete AE, when signals do not overlap, and we obtain the distribution of the pulse amplitudes. For continuous AE, the amplitude distribution is different. Figure 1.8 shows the amplitude distribution of AE signals obtained during fracture of cylindrical samples of alloyed steel (0.38C–0.18V–0.35Si–0.3Cu–0.35Mo–0.25Mn–3Ni–0.025P–1.2Cr–0.025S wt%) [88].

Further studies of materials fracture allow us to make a conclusion that the frequency spectrum of AE signals, which is obtained by Fourier transform, gives important information about physical characteristics of material. In particular, it is possible to establish the duration of the source action, analyze the fine frequency structure of AE signal, separate the contribution of deformation and fracture to AE generation, distinguish useful signals from noise processes, and so on.

A large number of publications have been devoted to the study of the spectral characteristics of AE signals [89–107]. In particular, in [90], the features of the spectra of AE signals that are generated during the plastic zone growth are studied. Since in this case the highest point of the spectrum is practically unchanged, the authors argue that the increase in the total count of AE signals under failure is the result of an increase in the number of events, not the amplitude. Based on the frequency analysis of AE signals generated during fracture of HF-1 steel and 70–30 brass, it is suggested that the change in the signals frequency spectrum may be associated

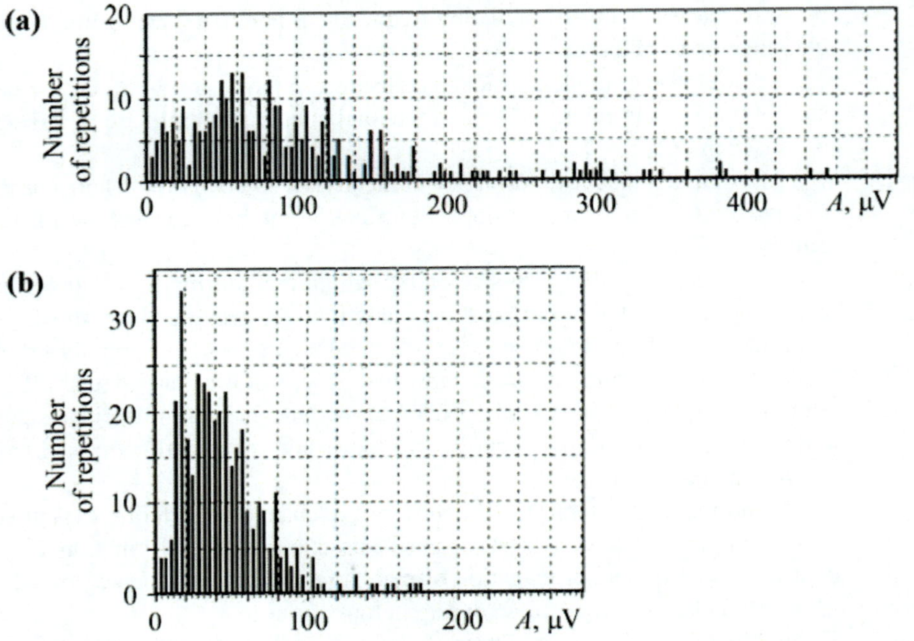

Fig. 1.8 Amplitude distribution of AE signals under brittle (**a**) and ductile (**b**) fracture of material

with the change in the deformation mechanisms [91]. In [92], it is established that with increasing damage (micro- and macrocracking) of material, the shape of the frequency spectrum of AE signals changes.

To identify AE signals from dislocations, the authors of [93, 94] investigated commercially pure aluminum and carbon steel (0.45C–0.17Si–0.25Cu–0.08As–0.8Mn–0.25Ni–0.035P–0.25Cr–0.04S wt%) as well as other materials—from microcracks. It is found that AE, caused by the movement of dislocations, quickly attenuates in a solid, thus indicating a high frequency of such signals.

Studies of the relationship between different mechanisms of AE sources and duration of the signals generated by them have shown that the spectrum of a shorter pulse is longer [95].

The authors of [96], studying the emission of low-carbon steel, have found that the frequency spectrum of the initial AE is within 500 kHz–3 MHz and characterizes the plastic deformation of the material. In investigations of the plastic deformation of aluminum monocrystals [97] using AE, they propose a method for estimating the instantaneous speed of dislocations, their free path length, and effective stress.

The correlation of the frequency spectra and different stages of plastic deformation with physical phenomena is established in [98]. Two regions of plastic deformation are distinguished: uniform elongation to the yield strength and a neck with their characteristic patterns. The pattern in the region of homogeneous elongation shows that during continuous AE, the dominant frequencies correspond to the motion of dislocations. The lower value of the dominant frequency is observed in the neck region, which probably corresponds to the coalescence of micropores and/or the microcracks growth.

The results of the investigation of plastic deformation of metals by the AE method are given in [99]. In particular, for carbon steel the highest intensity of AE at the moment of reaching the yield strength with a further sharp decrease in the plastic deformation zone is observed. The frequency spectrum of the signals is wide in the area of elastic and at the beginning of plastic deformation. It does not depend on the loading rate. In the zone of plastic deformations and up to the fracture, the spectrum narrows.

In order to recognize AE signals from non-critical sources (nucleation of voids under plastic deformation near a crack) and a growing crack, the frequency spectra of AE radiation are analyzed [100]. For AISI-4340 steel, the frequency peak of the signals is found in the range of more than 200 kHz. It is assumed that such AE signals correspond to the crack propagation. For SA 533B ductile steel, in the initial stages of the defect development, the signals have significant amplitudes in the high-frequency region. Under further loading, the shape of the spectrum is gradually shifting. At the end of the fracture, signals with a dominating low-frequency component appear. It has been suggested that such changes in the spectrum are caused by signals from other sources, in particular, from the voids nucleation under plastic deformation. The beginning of the stable crack growth is clearly determined by a sharp change in the slope of the curve of the dependence "total energy of AE radiation versus J-integral (J_{IC})".

Studying the peculiarities of deformation of siliceous iron, the researchers, based on the frequency analysis, identify the following stages of deformation: generation of signals with small amplitude and frequency indicates the early stages of loading, sliding; the appearance of high-frequency harmonics corresponds to prefracture, plastic yield of the material; further disappearance of high frequencies is associated with the development of fracture [101].

Based on the physical representation of macroscopic damages as statistical sets of elementary defects, the authors of [102] propose a model of formation of a collective AE source and an algorithm for secondary processing of its signals (according to differences in their spectra), which allow us to detect and recognize defects.

Theoretical calculations and results of experimental measurements of AE for two types of fracture (plastic deformation and crack propagation) based on the analysis of the frequency spectrum are presented in [103].

It is determined in [104] that the bandwidth between adjacent zeros of the spectral density of the AE pulse is inversely proportional to the crack length or size of the crack jump.

Aluminum alloy is investigated by the AE method under tension and fatigue to identify the crack initiation in [105]. The optical methods (infrared tomography and digital image correlation) are used to verify the results, and it is found that the AE signals that accompany crack initiation have a peak frequency exceeding 500 kHz. New parameters are proposed to determine the damage of material.

Investigation of low-carbon steels has indicated [106] that during deformation, AE signals with a frequency peak of less than 50 kHz are generated, and normal tear fracture (transcrystallite) is accompanied by signals with a frequency peak in the range of 200–500 kHz. Optical and scanning electron micrography is used to verify the obtained results.

The authors of [107] based on the results of theoretical and experimental studies constructed the dependence of the bandwidth on the stress intensity factor (SIF). It is established that the narrowing of the frequency band can be a criterion for identifying the stages of subcritical crack growth, using its width the SIF can be determined, and therefore, the strength of the structural element under operation can be calculated.

Improvement of the technological basis of AE equipment and expansion of the methods of processing the registered AE signals to identify types of failure in different materials (steels, alloys, composites) make it possible to use multiparametric criteria that improves the efficiency of identification.

AE from the growth of a crack and other sources is investigated in steel pressure vessels under loading [108]. It is found that brittle fracture is the main mechanism both during crack propagation and oxide cracking of steel. To identify the fracture type, different waveforms of AE signals are analyzed.

The role of dislocation sliding and twinning as deformation mechanisms in a magnesium alloy and the influence of the grain size on their mutual contribution are investigated in the publication [109]. To analyze AE signals, a new method of their classification is used. It is found that dislocation sliding and twinning occur during tensile loading simultaneously for both types of material microstructure (grain size 70 and 2 μm), although the transition from one prevailing mechanism to another is different. In particular, in fine-grained material, the plastic deformation is first caused by the sliding of dislocations, and then the mechanism of deformation twinning prevails. For a coarse-grained alloy, this sequence is inverse.

The AE method is also efficient for identifying other fracture mechanisms. For example, in [110], the results of monitoring by this method of physical processes associated with local electrochemical one, which occur in the electrodes ($MmNi_{3.6}Mn_{0.4}Al_{0.3}Co_{0.7}$ and $TiCr_{0.3}V_{1.8}Ni_{0.3}$), are presented. For this purpose, the AE signals are analyzed according to the following parameters: wave reflection, spectral power, and duration. When the positive electrode $MmNi_{3.6}Mn_{0.4}Al_{0.3}Co_{0.7}$ is charged, intensive cracking (a large number of small cracks) occurs due to hydrogen absorption, and for the negative electrode, $TiCr_{0.3}V_{1.8}Ni_{0.3}$ cracking starts after hydrogen evolution (larger cracks).

Using the analysis of waveforms and features of the AE signals frequency spectrum, the authors of [111] identify such wear mechanisms as adhesive and abrasive mechanical ones. It is found that the AE signal recorded during adhesive wear has a large peak in the high-frequency range and in abrasive one—several peaks in the low-frequency range.

The mechanisms of contact fatigue crack propagation are identified in [112]. Comprehensive analysis of AE signals (multiparametric analysis, location of AE signals during cyclic loading and statistical analysis) makes it possible to identify three different stages during crack propagation, which reflect the evolution of shear and tensile effects: (1) Mode II crack propagation (sliding mode crack); (2) mixed-mode crack propagation; (3) Mode I crack propagation (opening mode crack).

For a deeper understanding of the mechanisms of damage (fracture) of austenitic steel for nuclear components under cyclic loading, the conditional and multiparametric statistical analyses of AE signals are used [113]. As a result, the signals from plastic deformation and fatigue crack growth are identified.

1.2.3 Identification of Fracture Mechanisms of Composite Materials

Composite materials (CM) occupy an important place in many fields of industry due to their physical and mechanical properties, which provide the opportunity to combine the industrial design and the reliability of products.

Particular attention has been devoted to the study of the strength characteristics of fiber CM [114–116]; this is determined not only by the features and variety of their constituting elements but also the features of the materials structures. Series of the possible stages of macrofracture are emphasized for these composites, in particular, matrix cracking, breaking of adhesion, delayering of fibers from a matrix, fracture of fibers, their pullout, and so on. The AE method is widely used to study these processes [21, 117–141]. Let as consider some of them. In particular, in [130], during tensile loading of a metallized fiber-glass composite, AE signals are analyzed by pattern recognition and frequency analysis method. A correlation is established between the accumulated absolute AE energy and the mechanical properties of the samples with and without the coating under tension. It is found that a stronger adhesion to the coating substrate leads to a decrease in mechanical energy, which is emitted during elastic deformation and which can be recorded by equipment. Besides, signals from four fracture mechanisms of samples with a coating and three mechanisms of samples fracture without it are classified according to differences in frequency parameters and using the method of image identification. In the first case, it is cracking of the matrix, the fiber/matrix debonding, rupture of the fibers, and the matrix/coating debonding. The results of optical studies of samples as well as their surfaces of fracture (microfractography) confirmed the mechanisms identified by the AE analysis.

Two mechanisms of fracture associated with matrix cracking and delamination are distinguished in [131] using the analysis of AE signals by the method of neural networks during compression of the laminate with carbon fiber.

AE is used also to study the signal properties of individual concrete components (filler, matrix, and interfacial transition area) [132]. It is found that for cracks propagating in the filler, the frequency range of AE signals is 300–400 kHz when they develop in the matrix, and in the interfacial area, it is 100–300 kHz.

Various fracture mechanisms in multilayer piezoelectric composites have been studied according to the dominant frequency band of AE signals [133]. It is found that during catastrophic failure in the PZT-layers for AE signals, the frequency band of 150–250 kHz with large amplitudes is dominant. For cracking of the matrix and/or delamination between the fiber and the matrix, the frequency band is 150–250 kHz with their small values. During fracture of the fiber and/or delamination between the fiber and the PZT-layer, the frequency band is 300–500 kHz with high amplitudes of AE signals.

The characteristics of fracture for aramid fiber-reinforced composite under the influence of high impact force are studied in [134]. AE signals are analyzed by parameters such as amplitude, normalized total energy, ratio of signal rise time to its amplitude, and frequency peak by fast Fourier transform (FFT). Four types of signals according to the signal frequency peak and its value that accompanied the corresponding failure mechanism are identified. They are matrix cracking with low level of amplitudes (42–48 dB) and frequency peak (about 106 kHz); fiber breakage—high level of amplitudes (more than 70 dB) and frequency peak (more than 300 kHz); fiber/matrix debonding and fiber pullout—medium level amplitudes (50–65 dB) and low frequencies (about 110 kHz); fiber splitting and fibrillation—low amplitudes (below 55 dB), but high frequencies (about 400 kHz).

Fracture of laminate composites with different configurations of fiber packing and different fiber systems is studied in [135]. After classifying AE signals by their dominant frequencies and amplitudes using FFT and performing microscopic studies, it is found that the direction of the main crack propagation in such composites depends significantly on the orientation of the fibers and does not depend on the direction of the initial cut on the sample.

According to the AE parameters (total number of AE events, counting rate, and frequency), the micromechanisms of fracture of three composite materials with different matrices (metal, polymer, ceramics) are identified in [136]. The AE results illustrate the different damage accumulation profiles and indicate the change in the fracture mechanisms during loading under the influence of each parameter (temperature, orientation, and density) for a particular composite. For example, for graphite-epoxy composites, according to the AE results, different fracture mechanisms in time are established: from fiber breakage with subsequent delamination to matrix cracking.

When studying the fracture of ceramic composites with carbon fibers, it is found that the development of their damage can be monitored by the total energy of AE signals [137]. Using the results of AE and microscopic studies, three stages of damage accumulation in the composite under tension are established. At stresses that make up about 40% of the fracture stress, damage of composites occurs due to the growth of microcracks in the matrix. At the second stage, many microcracks in the matrix begin to expand and coalesce to form a macrocrack that propagates through the carbon fibers that bind (keep) the composite. The fibers stretched in the direction of the load hold the fractured matrix, preventing a sharp rupture of the composite.

Applying the method of pattern recognition to AE signals under four-point bending, different mechanisms of fracture in the metal coating of composites with carbon fibers are identified in [138].

In the investigations of the fracture of composites with carbon fibers under tension and compact compression and tension, using the method of pattern recognition and frequency analysis of AE signals, each mechanism of their fracture is classified [139]. In particular, during the matrix cracking, the frequency of AE signal ranges 1–50 kHz, under the composite delamination the band is 50–150 kHz, the fiber/matrix debonding—200–300 kHz, fiber breakage—400–500 kHz, and fiber pulling from matrix—500–600 kHz. The obtained results correlate with the known ones [140, 141].

According to the analysis of literature sources to identify the fracture mechanisms of CM by the frequency parameters of AE signals, FFT is used, but in further studies WT is more often applied as an effective tool of time–frequency localization of signals generated under particular mechanism of composite fracture.

1.3 Detection of Defects by Signals of Magnetoelastic Acoustic Emission

To implement traditionally the AE method, to detect operational defects such as cracks, it is necessary to apply an external load to the object of control. It may happen that the load under which small defects propagate will significantly exceed the optimal modes for the object of study. In this case, for the local diagnostics of ferromagnetic elements of structures and products, the phenomenon of generating the signals of magnetoelastic acoustic emission (MAE) under the influence of an external magnetic field, which causes the movement of the walls of magnetic domains, is used. It is possible to develop such approach for not only efficient detection of defects, but also for quantitative evaluation of the degree of degradation of structural materials by combining the parameters of the MAE signals with known physical phenomena.

The MAE method is intensively developed for NDT of ferromagnetic materials [142, 143]. It is characterized by a large number of informative parameters and is associated with the replacement of the domain structure. The results of studies of the influence of many factors on the MAE signal parameters such as residual stresses [144–146], microstructural features [142–149] and microhardness [148] of ferromagnetic materials, the frequency of remagnetization of the external magnetic field [150–153], heat treatment modes [154, 155], plastic deformation [156, 157], and other factors inherent in structural materials are illustrated well enough in the literature. The works are known in which the effect of demagnetization and random fields [158], anisotropy [159], irradiation [160], hydrogen [22], etc., are studied.

The frequency spectrum of signals belongs to the less studied informative parameters of MAE. The authors of [161] have found that the amplitude and frequency of the MAE depend on the shape of the sample and the mode of its processing. In [162], the influence of the remagnetization frequency and the sample thickness on the parameters of the MAE signals for different steel grades has been studied, taking into account the FFT spectra. It is found that the geometric dimensions of the samples, and not the frequency of remagnetization, have a decisive influence on the frequency parameters of the MAE signals. In particular, for thin samples with a thickness of $h = 3$ mm, low-frequency components are practically absent, and the dominant frequencies of the MAE signals are concentrated in the range of 200–300 kHz. In the other case, for $h = 10$ mm, two predominant frequency ranges of 100–200 and 200–300 kHz are obtained. The spectra do not differ qualitatively for triangular pulse with different values of the remagnetization frequency (0.28 and 13.1 Hz).

The MAE of the samples of various iron-containing meteorites and their artificial analogs (two types of alloys: Fe, Ni-05.26%, Co-0.74% which was cooled to room temperature in air and Fe, Ni-5.17%, Co-0.66% which was cooled using furnace) are studied in [163]. If the MAE curves of meteorites and manufactured alloys belong to the same type due to the similar sample composition, the signal amplitudes, location, and width of peaks as well as the character of the MAE frequency spectrum have significant differences due to internal structure and magnetic texture. There are two peaks in the frequency spectra of the MAE of Fe (Ni, Co) alloys at frequencies of 70–80 and 100 kHz. The MAE signals in the case of meteorites have the same specific feature, but these peaks do not stand out as much, or absent at all. The behavior of the MAE signals of Fe (Ni, Co) alloys differs depending on the cooling conditions. In both cases, the curves have the same shape with two flat maxima located in the same range. However, for the alloy cooled in the furnace, the MAE signal is approximately twice as high as for the sample cooled in air. Shifts in the MAE signals spectrum maxima toward lower frequencies are also observed.

The average value of the envelope of elastic oscillations flow does not allow us to estimate the sources of the MAE selectively and determine their relationship to the structure of the material. The MAE spectrum in various alloys is investigated using a broadband AE converter and digital signal processing [164]. Two groups of samples are selected: The first one includes samples of iron-based alloys (Permendur, armco iron, steel with different carbon content); the second series includes the heat-treated samples of three structural steels with different carbon content. For a few alloys with different magnetic and magnetostrictive properties, it is found that the amplitude of the harmonic with a frequency equal to twice the frequency of the remagnetizing field correlates with the magnetostrictive sensitivity of material. The main frequency of the MAE changes slightly with increasing steel tempering temperature to 350–400 °C and increases intensively with its further growth. It is also found that the MAE amplitude–frequency parameters of the studied steels correlate with their residual magnetic induction.

The frequency dependence of the asymmetry of the alternating magnetic field in the range 1–200 Hz is studied in [165]. The correlation between the amplitude dependence of the MAE asymmetry and the strength of the interaction of domain wall defects in iron-based ferromagnets has been previously proved. Low-carbon steel samples with three levels of interaction strength of domain wall defects are tested. It is found that with increasing strength of the interaction, there is a vertical shift down of the asymmetry in the entire studied frequency range.

By literature review, the MAE signals are analyzed using the dependence of some parameter of signals (e.g., amplitude or intensity) on some properties of the object and the experimental conditions. For higher efficiency of the MAE method, it is possible to improve the methods of signal processing.

1.4 Methods of Spectral Analysis of AE Signals

The AE signal processing is based on the spectral analysis of events as average characteristics of various parameters obtained by their statistical analysis. Traditionally, the Fourier transform is used to evaluate the signal properties in the frequency domain [166].

Note that AE in structural materials is a stochastic process that depends on the physical nature and random character of internal stresses at local points of the materials. As a result, the AE energy can be represented as the sum from its various sources containing the background noise. Thus, the AE spectra are more complex characteristics, that include reflections from the local and remote energy sources, as well as various mechanisms of acoustic signal formation.

During the materials fracture, elementary acts can have different durations and overlap in time. For the analysis of structure and behavior of the signals, that contain the domains of various duration with fluctuations of various speeds or frequency filling, the WT is used. The signal with time-localized areas of different frequency properties is decomposed into a set of wavelet functions, which are also localized in time and have differing spectral composition. Such decomposition guarantees local analysis of the signal: If a certain expansion factor is of great importance, then setting the time interval to which it corresponds, one can analyze it in detail, determines the frequency composition and duration. There is almost no alternative to WT for the investigation of irregularities in the signals. A theory of WT is presented in detail in many sources, in particular, in [167–171]. Traditionally, there are two types of WT: continuous and discrete.

1.4.1 Continuous Wavelet Transform

Consider a continuous signal of infinite duration, which is described by the function $s(t)$. This function is a vector of the Hilbert space of square-integrable continuous functions of one variable (with respect to Lebesgue measure): $s(t) \in L^2(R)$, where R is the set of real numbers. The scalar product of two vectors is set in the space of existence of signals:

$$\langle f(t), s(t) \rangle = \int_{-\infty}^{\infty} f(t) s^*(t) \mathrm{d}t,$$

where the symbol * indicates a complex conjugation.

The norm is set for signals

$$\|s(t)\| = \sqrt{\int_{-\infty}^{\infty} |s(t)|^2 \mathrm{d}t}.$$

Fourier transform of a function is

$$\tilde{f}(\omega) = \int_{-\infty}^{\infty} f(t) e^{i\omega t} \mathrm{d}t.$$

A real-valid function $\psi(t) \in L^2(R)$ that satisfies the existence condition

$$C_\psi = \int_0^{\infty} \frac{\left|\tilde{\psi}(\omega)\right|^2}{\omega} \mathrm{d}\omega < \infty \tag{1.1}$$

and has a unit norm $\|\psi(t)\| = 1$, is called the mother wavelet. For a given scale a and shear b, this wavelet generates a set of extended or compressed wavelet functions, each of which can be obtained by the rule:

$$\psi_{ab}(t) = \frac{1}{\sqrt{a}} \psi\left(\frac{t-b}{a}\right). \tag{1.2}$$

If the mother wavelet is chosen and a set of functions (1.2) is formed, the continuous wavelet transform (CWT) of the continuous function $s(t)$ with respect to the mother function $\psi(t)$ is the function in the form

$$W(a, b) = \langle s(t), \psi_{a,b}(t) \rangle = \frac{1}{\sqrt{a}} \int_{-\infty}^{\infty} s(t) \psi^*\left(\frac{t-b}{a}\right) \mathrm{d}t, \; a \neq 0. \tag{1.3}$$

$W(a, b)$ is a continuous function of two real variables: the scale $a \in R$, $a \neq 0$ and the shear $b \in R$. Very often in practice, the scale is limited to values $a \geq 0$ (condition (1.1) is set for such a case).

Fulfillment of condition (1.1) is equivalent to the requirement of zero equality of the mother wavelet average value [167]:

$$\int_{-\infty}^{\infty} \psi(t) \mathrm{d}t = 0.$$

According to the Calderon–Grossman–Morlet theorem, if the existence condition (1.1) is fulfilled, any function $s(t) \in L^2(R)$ satisfies the equalities

$$s(t) = \frac{1}{C_\psi} \int_{-\infty}^{\infty} \int_{-\infty}^{\infty} W(a, b)\psi_{ab}(t) \frac{da db}{a^2},$$

$$\int_{-\infty}^{\infty} |s(t)|^2 dt = \frac{1}{C_\psi} \int_{0}^{\infty} \int_{-\infty}^{\infty} |W(a, b)^2| \frac{da db}{a^2},$$

and consequently, the direct CWT has the inverse one and is a transform with energy conservation.

Therefore, the convolution of a signal with one of the wavelets provides the extraction of characteristic features of the signal in the area of this wavelet localization. A wider area of the signal has an impact on the convolution results for a wavelet of larger scale.

According to the uncertainty principle, the better the concentration of the function in the time domain, the more spread it will be in the frequency domain. Due to the rescaling of the function, the product of the time and frequency regions remains constant and is equal to the area of the cell in the frequency–time (phase) plane. The advantage of the WT is that it covers the phase plane with cells of the same area but of different shapes (Fig. 1.9). This facilitates the localization of low-frequency signal details in the frequency domain (predominant harmonics) and high-frequency details in the time domain (sharp jumps, peaks, etc.).

Up-to-date processing makes it possible to present AE signals in the time and frequency domains as one graphical representation of WT. During the practical analysis of frequency inhomogeneities in the signal, the timescale of the wavelet spectrum is somewhat unusual for visual perception, so it is replaced by a frequency scale [172].

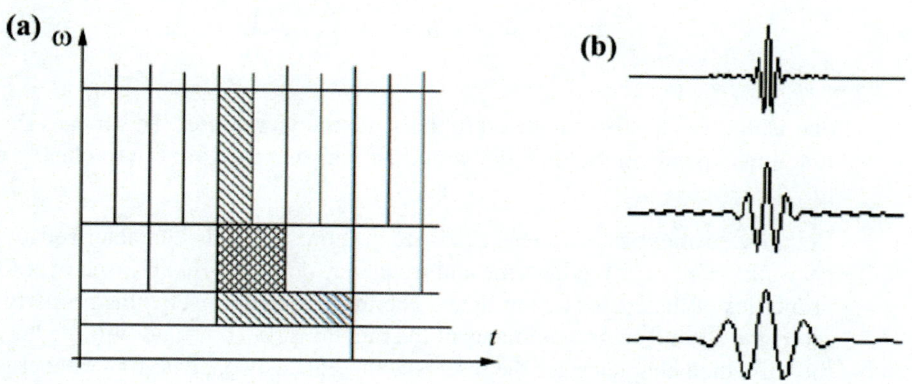

Fig. 1.9 Schematic representation of WT on the plane ω–t (**a**) with basic wavelet functions of different scales ($a = 2^k$, $k = 0, 1, 2$) (**b**)

To pass to the last one, they determine the average wavelet frequency f_0 on a unitary scale ($a = 1$). This can be done directly in the time domain for the maximum of the cross-correlation function:

$$p(f) = \int_{-\infty}^{\infty} \psi(t, 1, 0) \cdot \cos(2\pi f t)\mathrm{d}t.$$

After determination of f_0, the frequency scale is calculated by the timescale using transformation $f_0/a \to f$.

Direct WT contains a combined information about the signal and the wavelet. At the same time, it allows us to obtain the objective information about the signal, since some WT properties do not depend on the choice of the wavelet. Owing to the wavelet independence, these simple properties are very important. Below we recall main of them.

Linearity. The linearity of the CWT follows from the scalar product (1.3):

$$W[\alpha f_1(t) + \beta f_2(t)] = \alpha W_1(a, b) + \beta W_2(a, b).$$

Shift. Consider the CWT function $f_1(t) = f(t - b')$. Then

$$W_{f_1}(a, b) = W_f\big(a, b - b'\big).$$

that is, the signal shifting in the time domain by a magnitude b' leads to shifting the wavelet image by the same magnitude.

Scaling. Consider the CWT of the function $f_1(t) = (1/\sqrt{c}) f(t/c)$, where the multiplier $1/\sqrt{c}$ is introduced for energy conservation. After substitution, we have

$$W_{f_1}(a, b) = W_f\left(\frac{a}{c}, \frac{b}{c}\right),$$

that is, the WT is also subjected to the scale transformation. This means that if the function expands in the time domain, then it also expands in the frequency domain [*a*, *b*].

Locality properties. The energy concentration, which is characterized by of the basis function $\psi_{ab}(t)$ in the time and frequency domains, is expressed over the local properties of the basis (1.2) in them. Therefore, quantitatively, this property can be set by the second central moments of the functions $|\psi_{ab}(t)|^2$ and $|\Psi(\omega)|^2$, the square root of which characterizes the region of the greatest concentration of energy given by these functions.

Denote by t_0 and Δ_t^2 the values of the first and second central moments of the function $|\psi(t)|^2$, i.e.,

$$t_0 = \frac{1}{\|\psi\|^2} \int_{-\infty}^{\infty} t|\psi(t)|^2 \mathrm{d}t, \ \Delta_1^2 = \frac{1}{\|\psi\|^2} \left[\int_{-\infty}^{\infty} (t - t_0)^2 |\psi(t)|^2 \mathrm{d}t \right]$$

Then the first moment (center of gravity, average value of the function) $|\psi_{ab}(t)|^2$ will be equal to:

$$\bar{t}_1 = \frac{1}{a\|\psi\|^2} \int_{-\infty}^{\infty} t \left|\psi\left(\frac{t-b}{a}\right)\right|^2 \mathrm{d}t.$$

Substituting the variable $t_1 = \frac{t-b}{a}$ and making some transforms, we get

$$\bar{t}_1 = \frac{a}{\|\psi\|^2} \int_{-\infty}^{\infty} t_1 |\psi(t_1)|^2 \mathrm{d}t_1 + b \frac{1}{\|\psi\|^2} \int_{-\infty}^{\infty} |\psi(t_1)|^2 \mathrm{d}t_1 = b + at_0.$$

It follows that the center of gravity of the basis function $\psi_{ab}(t)$ changes with the scale a: the wider the basis function (larger the scale a), the greater the distance between the centers of the basis functions.

The second central moment $|\psi_{ab}(t)|^2$ will be

$$\Delta_\psi^2 = \frac{a^2}{\|\psi\|^2} \left[\int_{-\infty}^{\infty} (t_1 - t_0)^2 |\psi(t_1)|^2 \mathrm{d}t_1 \right] = a^2 \Delta_t^2.$$

Then the width (radius) of the function $\psi_{ab}(t)$ in the time domain $R\psi = a\Delta_t$, and the time "window" that it occupies in this domain around the point $t_w = b + at_0$, $\mathrm{Win}_t = [b + at_0 - a\Delta t, b + at_0 + a\Delta_t]$. This means, on the one hand, that the value of the analyzed function $f(t)$ at some point t_0 affects the value of the coefficient $W_f(a, b)$ with increasing scale over an increasing period of time, forming an angle of influence in the plane (a, b) (Fig. 1.10a). On the other hand, the value of the coefficient $W_f(a, b)$ at point (a_0, b_0) will be determined by the value of the function $f(t)$ the vicinity of point b_0, which is found at the same angle of influence (Fig. 1.10b).

An important conclusion follows from Fig. 1.10 that small-scale variations of function $f(t)$ can be detected for small-scale values a, while for large values they will be smoothed out, but in this case the large-scale changes of this function can be found. Function $\Psi(\omega)$ describes a bandpass filter with a central frequency ω_0 due to the zero value of the first moments of the wavelet $\psi(t)$.

If we introduce function $\eta(\omega) = \Psi(\omega - \omega_0)$, that has a zero mean value and width Δ_ω, then WT in the Fourier domain can be written as

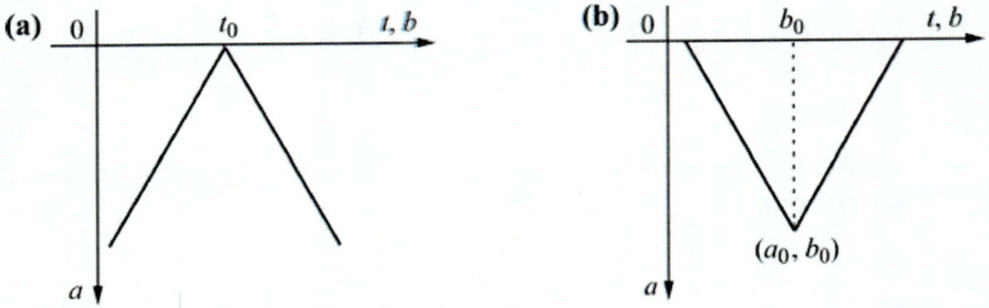

Fig. 1.10 Angle of influence of function $f(t)$ values: **a** on the coefficients $W_f(a,\ b)$ with increasing scale; **b** on the coefficient $W_f(a_0,\ b_0)$ at point b_0

$$W_f(a,b) = \frac{a\sqrt{a}}{2\pi}\int_{-\infty}^{\infty} \exp\{ib\omega\}\eta\Big(a\Big(\omega - \frac{\omega_0}{a}\Big)\Big)F(\omega)\mathrm{d}\omega.$$

The window function $\eta\big(a\big(\omega - \frac{\omega_0}{a}\big)\big) = \eta(a\omega - \omega_0) = \Psi(a\omega)$ has a width ω_0/a.

Therefore, up to a constant and phase shift $\exp\{ib\omega\}$, WT gives some local information about the spectrum of the analyzed function $F(\omega)$ in the band $\pm\Delta_\omega/a$ with a center at point ω_0/a, i.e., in the window $\mathrm{Win}_\omega = \big[\frac{\omega_0}{a} - \frac{\Delta_\omega}{a}, \frac{\omega_0}{a} + \frac{\Delta_\omega}{a}\big]$.

At the same time, the ratio of the center frequency ω_0/a to the width of the window $2\Delta_\omega/a$ remains constant at all a values. As the scale increases (shift to the low-frequency range of the analysis), the window will expand along the time axis in the time–frequency plane (Fig. 1.10) and will narrow along the frequency scale. When the scale decreases, it is vice versa.

1.4.2 Discrete Wavelet Transform

In the discrete domain, the parameters of scale a and shift b are discretized as $a = a_0^m$ and $b = nb_0$. Then the wavelets take the form

$$\psi_{m,n}(t) = a_0^{-m/2}\psi\left(\frac{t - nb_0}{a_0^m}\right), m, n \in Z,$$

where m and n are the scale and shift parameters, respectively. The discrete wavelet transform (DWT) and its inverse transform can be written as follows:

$$W_{m,n} = \int_{-\infty}^{\infty} s(t)\psi_{m,n}^*(t)\mathrm{d}t$$

$$s(t) = k_\psi \sum_m \sum_n W_{m,n}\psi_{m,n}(t),$$

where k_ψ is the normalizing coefficient.

In general, there can be any value of the scale a, but usually it is $a = 2$, and the transform is called as dyadic WT. A fast computational algorithm called multiple-scale analysis (MSA) has been developed for the dyadic WT. The number of wavelets according to the scale factor m determines the level of signal decomposition, while as the zero level ($m = 0$) the level of maximum time division of the signal is often taken, i.e., the signal itself, and the following levels form a descending wavelet tree.

According to the MSA concept, the signal $s(t)$ is decomposed into two components: approximating $A_1(t)$ and detailing $D_1(t)$:

$$s(t) = A_1(t) + D_1(t) = \sum_k a_{1k}\varphi_{1k}(t) + \sum_k d_{1k}\psi_{1k}.$$

Two new sequences a_{1k} and d_{1k} are obtained, the length of each is equal to half the length of a_{0k}. Then the expansion is continued by $A_1(t)$. At the level of decomposition m, the signal $s(t)$ is written as

$$s(t) = A_m(t) + \sum_{j=1}^{m} D_j(t) = \sum_k a_{mk}\varphi_{mk}(t) + \sum_{j=1}^{m}\sum_k d_{jk}\psi_{jk}(t), \tag{1.4}$$

where $a_{mk} = (s(t), \varphi_{mk}(t))$ and $d_{jk} = (s(t), \psi_{jk}(t))$ are approximation and detailization coefficients, respectively.

As we see from relation (1.4), a_{mk} and d_{mk} depend on continuous basis functions $\varphi(t)$ and $\psi(t)$, which are unambiguously determined by coefficients h_l [167]:

$$\varphi(t) = 2\sum_l h_l\varphi(2t - l), \tag{1.5}$$

$$\psi(t) = 2\sum_l (-1)^l h_{2n-1-l}\varphi(2t - l) = 2\sum_l g_l\varphi(2t - l),$$

$$h_l = h(l) = (\varphi(t), \varphi(2t - l)), g_l = (-1)^l h_{2n-1-l}, \tag{1.6}$$

where $l = 2n - 1$ and n is the wavelet order.

In practice, the smallest possible scale (highest resolution n_0) determines the number of discrete values for signal ($N = 2^{n_0}$). For the smallest value of the scale ($m = 0,\ a = 2^m = 1$), the readings s_i of the signal $s(t)$ are taken as approximation coefficients a_{0k}, i.e., $a_{0k} = s_i, k = i,\ i = 0,\ 1,\ ...,\ N - 1$. From the representations (1.5) and (1.6), it is possible to obtain recurrent relations for the fast wavelet analysis during the transition from the scale m to the next one $m + 1$ and

$$a_{m+1,k} = \sum_l h_{l-2k}a_{ml}, \quad d_{m+1,k} = \sum_l g_{l-2k}a_{ml}.$$

Here h_l, g_l are low- and high-frequency filters, respectively.

The process stops after reaching a finite number of levels, which depends on the signal length (N) and the order (l) of the filter h_l. Figure 1.11a shows a scheme of the MSA algorithm.

Most of the information about the output signal is concentrated mainly in its low-frequency region, the decomposition of which can be extended up to zero. As a result, a "one-sided" tree is obtained (Fig. 1.11a). However, for many real signals, the high-frequency region contains an important information. A similar operation can be applied to any part of the decomposition. This corresponds to the replacement of the wavelet with two new wavelets

$$\psi_1(t) = \sum_n h_n \psi(t-n), \psi_2(t) = \sum_n g_n \psi(t-n),$$

which are also localized in space, but in an interval twice as wide as the original one. The binary tree of the decomposition in Fig. 1.11a "splits" for coefficients "D" of any level. As a result, a "complete" (binary or balanced) tree is obtained in Fig. 1.11b. The branches of the tree correspond to the sets of signal spaces with bases constructed as for a one-sided tree, according to MSA.

The functions that generate these bases are called wavelet packets, and the signal decomposition is called a discrete packet wavelet transform (DPWT). If only the low-frequency part is decomposed in the DWT, both low- and high-frequency parts are decomposed in the DWPT. Thus, a packet WT promotes better frequency localization of the signals.

During the signal decomposition at the wavelet level, a certain frequency range corresponds to each of them. In particular, the frequency range for the signal decomposition into J levels is distributed by the rule $0 \sim \frac{1}{2^J} f_{\max}$, $\frac{1}{2^J} f_{\max} \sim \frac{2}{2^J} f_{\max}$, …, $\frac{2^J-1}{2^J} f_{\max} \sim f_{\max}$, totally 2^J wavelet components [173]. A certain mechanism of fracture (plastic deformation, growth of micro- and macrocracks, etc.) is characterized by AE signals with corresponding predominant frequency ranges. Since the energy distribution for each wavelet component is different, it is possible by applying energy analysis to the WT of these signals, to detect these frequency ranges, and thus identify the type or mechanism of fracture.

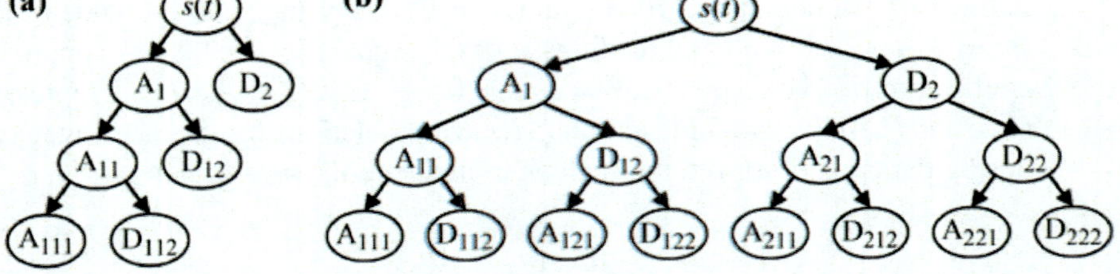

Fig. 1.11 Scheme of algorithms of a discrete (**a**) and discrete packet (**b**) WT

For the signal $s(t)$, let us denote the J-level components s_J^1, s_J^2, ..., s_J^i, of its wavelet decomposition and their energies as E_J^1, E_J^2, ..., E_J^i, respectively. Then the accumulated energy for each component of the signal can be mathematically written in the form

$$E_J^i(t) = \sum_{\tau=t_0}^{t} (s_J^i(\tau))^2,$$

and a total energy of signal looks like this

$$E_{\text{total}}(t) = \sum_J E_J^i(t). \tag{1.7}$$

Therefore, the energy distribution is considered as a ratio of the energy of each component of the decomposition to the total signal energy [174], i.e.,

$$rE_{\text{WT}} = \frac{E_J^i(t)}{E_{\text{total}}(t)}, \quad j = 1, 2, \ldots, 2^J, \tag{1.8}$$

Since the energy distribution, defined by expression (1.8), contains not only temporal information (duration and amplitude), but also frequency one, the value rE_{WT} can serve as an information parameter for recognizing AE signals. Using the wavelet algorithm, AE signals can be classified, for example, by defining their typical models.

Mother wavelets, on the basis of which the WT functional space is built, can be orthogonal, semi-orthogonal, biorthogonal; continuous, discrete, complex; symmetric, asymmetric, and non-symmetric. The wavelets are distinguished according to the degree of smoothness, with a compact area of definition and without it. Some functions have an analytical expression. A fast algorithm for calculating the associated WT is proposed for the others [175].

The choice of the wavelet depends on the type of information needed from the signal. Taking into account the characteristic features of different wavelets in the time and frequency domains, it is possible to detect certain properties and features in the signals that are not visible on the graphs of the signals, especially due to a noise. The most common real continuous bases are constructed on the basis of derivatives of the Gaussian function, because it has the best localization indices in both the time and frequency domains. Among discrete wavelets, Daubechies wavelets are the most popular, and among complex wavelets, the basis is most often used, which is based on the well-localized Morlet wavelet in both time and frequency domains.

1.5 Application of Wavelet Transform for Analysis of AE Signals

The AE method is one of a few methods of technical diagnostics and nondestructive testing that enables one to observe the initiation and development of defects in the materials of objects [176]. While monitoring the state of high-cost or potentially dangerous equipment, it is important to be able to evaluate the variations of values of the parameters of AE signals as temporal functions in real time. For this purpose, it is possible to use the WT of AE response, which enables one to analyze the fine structure of signals [177]. Applying the WT to signals that are non-stationary in time or inhomogeneous in the space, it is possible not only to determine their general frequency characteristic (distribution of the signal energy over the frequency components) but also to collect information about certain local coordinates characterized by the manifestation of one or another group of frequency components or their rapid changes.

1.5.1 Wavelet Analysis of AE Signals in the Course of TD

A mathematical model of the automated system for TD and prediction of the residual service life of industrial equipment on the basis of complex application of the wavelet analysis and neural network technologies is well known [178]. The wavelet analysis makes it possible to separate the noise component affecting the accuracy of predictions. As a result, it becomes possible to realize the early detection of deviations of the mechanical characteristics of the product from permissible values. By using the neural networks, one can automate the process of monitoring and, hence, substantially decrease the efficiency of diagnostic procedures.

The results of application of the WT analysis in AE monitoring the state of tools and machines can be found in the surveys [179–181]. The following trends of the WT application are distinguished: the time–frequency analysis of signals; removal of noise components from the AE response and extraction of weak signals; determination of their characteristics; singular analysis of the state of tool and/or machines; determination of the characteristics of detected damages and defects; evaluation of the degree of wear of the tool, compression of information, and identification of systems.

For the analysis of AE signals in the process of diagnostics of engineering systems and equipment, it is customary to use the DPWT. Thus, in particular, the DPWT is used to detect AE signals generated by generation and propagation of cracks, formation of scratches in the course of monitoring of the state of tribological systems [182, 183]. In the investigation of the contact fatigue of railroad tracks [182] by the proposed method of selection of individual AE events, the process of crack propagation is analyzed and classified. The energy criterion of the DWPT for AE signals is used in monitoring the coating of the surfaces of cast iron–steel tribocouples

[183]. The authors of [184] determine the optimal mother wavelet for the DWPT according to the Shannon entropy criterion and propose new methods for the detection of defects in high-speed railroad tracks according to the frequency components of the AE signals.

Rolling bearings are important structural elements of various machines. Their state is monitored in four stages: statistical analysis, diagnostics of defects, determination of their sizes, and prediction of the residual service life. In [185], the authors propose an efficient modified algorithm for processing AE signals obtained from the components of rolling bearings for different rates of performing operations, loading rates, and the sizes of defects. Their approach is based on the optimization of the Kurtosis relation and Shannon entropy for determination of the optimal wavelet and application of the DWPT.

The work [186] is devoted to the problem of increasing the functional safety of the cars by introducing new methods of diagnostics of the gear box. The authors combined the methods of AE with vibroacoustic measurements and performed a DWT for the analysis of the data. In this way, they identify the most sensitive diagnostic parameters by using them it is possible to characterize the state of the investigated system better than with the help of the classical approaches.

In [187], the DWT is used for the analysis of the frequency content of AE signals in monitoring the procedure of grinding. The authors have established the influence of the surface roughness and geometric parameters of specimens on the frequency parameters of AE signals. They proposed efficient methods for the prediction of the surface quality of products and evaluation of errors of grinding.

The authors of [188] developed a new approach to the classification of AE signals appearing in the course of corrosion monitoring. Its important step is the removal of the noise components of signals with the help of WT performed in order to select the useful part of these signals. Reliable and efficient criteria of corrosion detection in chemical industry are proposed according to the results of subsequent analysis of the AE signals with the help of the random forest algorithm.

1.5.2 WT in the Problems of AE Sources Location and Identification of Defects in Structural Materials

In the literature, the application of WT for the location of AE sources [189–200] and identification of defects in structural materials are described fairly comprehensively [105, 173, 174, 201–212].

For the CWT of AE signals, the wavelets based on the Gaussian function, e.g., the Gabor wavelet, are used as mother wavelets especially extensively. By using this approach, the authors of [193] improve the procedure of filtering out the noise components from the signals, visualize and separate acoustic modes, determine the distance from the AE source according to a single oscillogram, and find the time of arrival of the signal more precisely. The methods of location of the AE sources with one transducer in thin plates are constructed in [194]. The time of appearance of the characteristic modes is determined according to the maximum coefficients of the Gabor CWT. Such wavelet is used in [205] to identify and evaluate defects in corroded steel, which make it possible to determine the frequency ranges corresponding to some types of defects. The same methods are used in [208] for a more efficient identification of different mechanisms of fracture of the coatings (transverse cracking, fracture of the coating/matrix bonding). It is proposed to use a frequency that corresponds to the wavelet coefficient with maximum amplitude. For the localization of defects in reinforced concrete structures, the authors of [200] use a new energy parameter based on the CWT with the Gabor wavelet.

The investigations based on the use of some other wavelet functions are also known. Thus, in particular, in [206], the efficiency of the CWT is analyzed on the basis of three wavelets (Gaussian 27, Gaussian 1, and Daubechies 40). A method is described for the detection and precise identification of AE events in time and location of the sources of signals according to the outflow within the corresponding energy ranges with the help of the Gaussian 27 wavelet. A method intended for the identification of cracks in plates on the basis of the wavelet analysis is presented in [192]. The CWT method is applied to AE signals for the location and determination of crack depths after choosing (in advance) the most efficient mother wavelet, namely symmetric 4, from the collection of wavelets including Coiflet 2, orthogonal 6.8, and symmetric 4. In [197], in order to detect defects in a wind-turbine box, the authors propose a method for more precise determination of the time of arrival of AE signals based on the WT with the Morlet wavelet. The sequence of micromechanisms of fracture in galvanized steel is discovered according to the variations of energy of the CWT with the Haar wavelet [209].

The DWT of recorded signals is often used for the detection of AE sources. Thus, in [173], to identify the time of initiation of friction in the rotor, the authors first perform the DWT of AE signals with the Daubechies 20 wavelet (eight levels of decomposition) with removal of the noise components. Then the site of friction is found by the cross-correlation method. In [195], for the identification of fatigue crack at the onset of its growth, the authors apply the DWT with biorthogonal wavelet (five levels of decomposition) to the AE signals. The application of the AE method for the monitoring of thermal welding was studied in [207]. On the basis of the DWT with the Daubechies 6 wavelet (six levels of decomposition), it is established that the locations and sizes of welding defects can be reproduced by analyzing the variations of the energy levels of decomposition of AE signals obtained in the course of passing of the welder over the defective area.

For the detailed analysis of the frequency spectrum of AE signals, the DPWT proves to be more efficient. Thus, in particular, in [189], the DPWT with the Daubechies 5 wavelet and four levels of decomposition (sixteen wavelet packets) are used to study the propagation of AE signals through the elements of turbines. The authors of [199] use the DPWT for the localization of welding defects. The procedure of identification of the micromechanisms of fracture in two-phase steel is developed in [202]. On the basis of the analysis of the frequency range of each level of decomposition of the DPWT of AE signals with the Daubechies wavelet (three levels of decomposition and eight wavelet packets) and the distribution of energy, it becomes possible to identify various types of defects, such as breaking of ferritic/martensitic bonds and failures of the martensitic phase. A new method for monitoring the state of bearings of rotating machines with the help of DPWT based on the Daubechies 10 wavelet is described in [210]. In [211], a grinding scorch of the material is identified with the help of the AE method. The properties of AE in the course of thermal expansion are investigated by modeling the thermal consequences of grinding by laser irradiation. According to the results of DPWT, it is established that the distribution of energy of AE signals at a high temperature is concentrated in high-frequency bands. At low temperatures, their distribution is concentrated in low-frequency regions. In the presence of a grinding scorch, the high-frequency characteristics of AE signals are much more expressed than in its absence.

In [212], one can find a method of two-dimensional WT used for the identification of cracks in plates. The crack depths are quantitatively evaluated by the maximum values and the energy contents of wavelet coefficients of the WT with the Symlet 6 mother wavelet.

The initiation of cracks in aluminum alloys in tension and in the process of fatigue fracture is investigated in [105]. To determine the characteristic features of the AE signals generated in the course of initiation and subsequent growth of a crack, the authors analyze the fracture diagrams, the amplitudes, and the frequency contents of AE signals in their CWT and compare them with the results of numerical modeling.

1.5.3 *Identification of the Mechanisms of Composites Fracture by Using the WT of AE Signals*

The WT of AE signals provides the time localization of their features and, hence, guarantees the efficient identification of the main fracture mechanisms in composites: matrix cracking, fiber/matrix debonding, fiber breakage and pullout, etc.

In particular, the classical AE method and the WT of AE signals are used to analyze the specific features of fracture of model composites with carbon fiber in [213]. By using the FFT, it is shown that the AE signals with predominant frequency range less than 100 kHz are generated during matrix cracking. The signals with the predominant frequency range 200–300 kHz are generated in the case of fiber/matrix debonding, whereas the signals with the predominant frequency range 400–450 kHz accompany the fiber breakage. To clarify the sequence of fracture mechanisms of the composites and their duration, the authors apply the CWT based on the Gabor wavelet and analyze the microfractograms of the specimens. It is shown that the fracture is caused by the violation of the integrity of the fiber at the center of the cross section. As a result, the cracks rapidly propagate along the radial lines in the fiber and in the matrix and caused the smash of the composite. In the case of WT, the duration of the emitted AE signal (its maximum frequency is 20 kHz) corresponding to the matrix cracking is 35 μs, while in the case of fiber breaking (its maximum frequency is 450 kHz), it is 20 μs.

The results of investigation of symmetric and asymmetric cracking of the matrix in the middle and surface layers of the glass fiber-reinforced composites are presented in [214]. The AE signals generated in the course of cracking of the matrix in multilayer composites are classified by using the generalized results of their numerical modeling the DWT (the Daubechies 5 wavelet) of modeled signals, the experimentally obtained signals, and the results obtained by the least squares method.

The procedure of application of the wavelet analysis of AE signals for the investigation of the specific features of fracture processes in composite materials hardened by carbon fibers is described in [215, 216]. To evaluate the AE signals obtained in the course of fracture of composites, the authors used the DWT with 11 levels of decomposition and the Daubechies 20 wavelet. It is shown that the ninth level characterizes the energy emitted as a result of fracture of the fiber and that the center of the frequency spectrum is close to 300 kHz. Moreover, the eighth level characterizes the energy of fiber/matrix breaking with a center of the frequency spectrum close to 250 kHz. The other sources of signals with higher or lower frequencies and low energies are also detected. Among these sources, the seventh level (with a frequency center close to 110 kHz) is predominant, which characterizes the energy generated as a result of matrix cracking.

The model of determination of the group velocity as a function of frequency of the transverse mode of normal waves propagating in graphite-epoxy composites on the basis of the analysis of the Gabor wavelet and the examples of its application for the efficient location of the sources of emission of AE signals is described in [217, 218]. By using the frequency dependence of the arrival time of signals and the angular dependence of the group velocity, one can establish more precisely the location of the AE sources for unidirectional and quasi-isotropic large flat composite plates.

In [219], the authors study a more complicated case of presence of the several types of waves propagating in a layer of the composite material. They propose general methods for the automatic selection of waves in composites with different structures for the DPWT of AE signals. It is established that, in this way, it is possible to identify important types of waves, determine the times of their arrival to the primary transducer, and, hence, increase the accuracy of location of the emission sources.

A new algorithm for the location of AE sources and evaluation of the group velocity in composite structures is proposed in [220]. This algorithm is based on the difference between the times of arrival of Lamb transverse waves recorded by six primary transducers. The CWT with the complex Morlet wavelet is applied to the AE signals. According to this concept, it is possible to identify the time of arrival of the wave packet and the value of frequency corresponding to it according to the squared amplitude of the maximum wavelet coefficient. To verify the algorithm, two different composite structures, namely a quasi-isotropic carbon fiber-reinforced polymer and a sandwich panel, are investigated. The sites of location of the AE sources and the group velocity are determined. The maximum error of location is approximately equal to 2% for the quasi-isotropic panel made of a carbon fiber-reinforced plastic and almost 1% for the plate.

The work [221] is devoted to the application of the AE method for monitoring of the state of composite materials being subjected to abrasive wear. Among the mechanisms of wear of fiber-reinforced composite materials, the following mechanisms can be distinguished: fiber/matrix debonding, rupture of the fiber and displacement of its parts by shear forces, and the removal of particles of the matrix. For the identification of signals from different mechanisms of wear, the authors use the method of pattern recognition. To separate the signals caused by specific events in the presence of continuous AE, the authors use a tool for the CWT built in the PACShare Wavelets (PAC) software. According to the analysis of the frequency features, the low-frequency and long-term signals correspond to the fiber/matrix debonding. The high-frequency signals correspond to the rupture of the fiber, and the AE signals of the medium-frequency range correspond to the matrix cracking. The signals generated as a result of fiber breaking and pullout, as well as the hybrid response combining the tones of the former two types, are identified using of a neural network.

The method of passive diagnostics for determination of the frequency characteristics of AE signals in graphite-epoxy composites is described in [222]. By using the DWT (with the Daubechies 4 wavelet and three levels of decomposition), it is established that the frequency range 300–400 kHz corresponds to the matrix cracking, while the frequencies lower than 250 kHz correspond to fiber/matrix debonding. It is also established that, at the end of the fracture processes in these composites, the duration of AE events decreases, while the amplitudes of signals are three times higher than in the case of fiber/matrix debonding.

In [223], the authors present the results of application of the multifactor analysis and WT for the identification of the types of fracture of glass fiber-reinforced composite materials with different structures, namely of multilayer materials and cross-ply composites. For the CWT and DWT of AE signals with the Daubechies 10 wavelet (five levels of decomposition), the authors proposed new criteria according to which the fracture mechanisms can be identified more efficiently than by the method of clustering of AE signals.

A complex methodology of post-processing of the AE signals obtained in the course of testing of the composites is proposed in [224]. For this purpose, the authors use the FFT for the determination of the predominant frequency ranges, and the WT (the Daubechies 20 wavelet and six levels of decomposition) aimed at the identification of fracture mechanisms. It is established that about 50% of energy of the AE signals is concentrated on the level of decomposition with a frequency range of 300–400 kHz. Thus, it may correspond to the most widespread mechanism of fracture of the composite, i.e., to the fiber breakage.

The DPWT is used to evaluate the state of the drilling tool and to study the specific features of fracture of glass-phenolic polymeric composites in drilling [225]. The analysis of the amplitude–frequency characteristics of AE signals demonstrates that, as the number of drilled holes increases, the frequency and amplitude of signals change. This may serve as the indication of deterioration the state of the tool. As a criterion for its evaluation, according to the results of investigations, it is proposed to use the parameter computed according to the wavelet coefficients of the DPWT (four levels of decomposition).

In [226], it is proposed to determine the mass fractions of substances in composites with microcrystalline graphite by the AE and DPWT methods. The characteristics of the AE signals attenuation in composites with various mass fractions of graphite are investigated. The relationships between the attenuation coefficients of energy, the amplitudes of packets of wavelet functions, and the mass fractions are established by using the DPWT. It is discovered that, for higher contents of graphite, the rates of energy attenuation and the amplitudes of packets of wavelet functions are higher. The parameters of attenuation within the frequency range 125–171.85 kHz are most sensitive to the evaluation of mass fractions. The error in this case is equal to 1.8% as compared with 3.9% according to the output signal.

Carbon and glass fiber/epoxy composites are studied in [227]. To determine the flexural loading, the authors use various methods of processing of AE signals, namely WT, FFT, and the Choi–Williams transform. It is established that, under the conditions of loading by three-point bending, the WT proves to be the best method that can be used for processing the AE signals.

The fracture of polymeric composites with carbon fibers is investigated in [228]. The correlation between the identified images and the mechanisms of fracture of these composites is established by using a new method of pattern recognition based on the coefficients of the CWT, which makes it possible to classify defects in these composites. The authors of [229] propose to use the DWT in order to remove the noise components from the AE signals.

In [230], the authors construct an algorithm for the identification of defects in composite walls of high-pressure vessels based on analysis of the CWT of AE signals. As a result, the types of waves (symmetric and asymmetric Lamb waves) are selected, and their frequency bands are determined.

The specific features of delamination of multilayer fiber-reinforced composites being subjected to tensile forces are investigated in [231]. For the analysis and clusterization of the AE signals, the authors used the method of pattern recognition. Further, on the basis of the results of finite element modeling, it is established that the obtained three types of AE signals correspond to the following mechanisms of fracture of fiber-reinforced composites: matrix cracking, fiber/matrix debonding, and fiber breakage. This result is also confirmed by the CWT of the experimental and modeled AE signals.

1.5.4 Wavelet Analysis in the MAE Signals Processing

Note that the MAE signals, just as elastic AE waves, are described by using the theory of random processes. They are characterized by the frequency spectrum that varies with time. This is why, for the analysis of their frequency features, it is more efficient to use the WT. These investigations are not numerous, which is reflected by a small number of available publications. In particular, in [232], one can find the procedure of evaluation of internal mechanical stresses according to the WT of the MAE signals. It is established that the wavelet coefficients in the frequency–time plane describe the characteristic features of signals and, moreover, by analyzing the variations of these signals, it is possible to evaluate the stressed states of structural elements.

The dependences of the Barkhausen noise and MAE signals on plastic deformation in plates of Armco iron are investigated in [233]. To detect the changes in the properties of noise signals, the authors perform the multiparametric analysis (FFT, amplitude distribution of signals, and WT). The influence of plastic deformation on the rate of generation of MAE signals and their amplitudes is determined.

The specific features of MAE in a strip of 2605SC metallic glass are studied in [234] by the remagnetization of specimens in two perpendicular directions. By using the CWT with the Gabor wavelet, the authors discover the differences between the frequency compositions of the MAE signals depending on the direction of remagnetization relative to the longitudinal axis of the strip: in the case of the parallel direction, the amplitudes of MAE signals are higher, more concentrated in time, and have a more complex frequency structure than in the case of the perpendicular direction (Fig. 1.12). This can be explained by the difference between the volumes of remagnetization. The authors select two types of MAE signals by using which it is possible to estimate the homogeneity of the structure of the material.

In [235], by the method of wavelet analysis, the authors established the mean duration of noise (about 10^{-4} s) in a $Fe_{78}B_{12}Si_9Ni_9$ amorphous alloy. By using a new approach proposed in [236], the Barkhausen noise is processed with the help of fractal analysis based on the WT. The cracks formed in metallic ferromagnetic

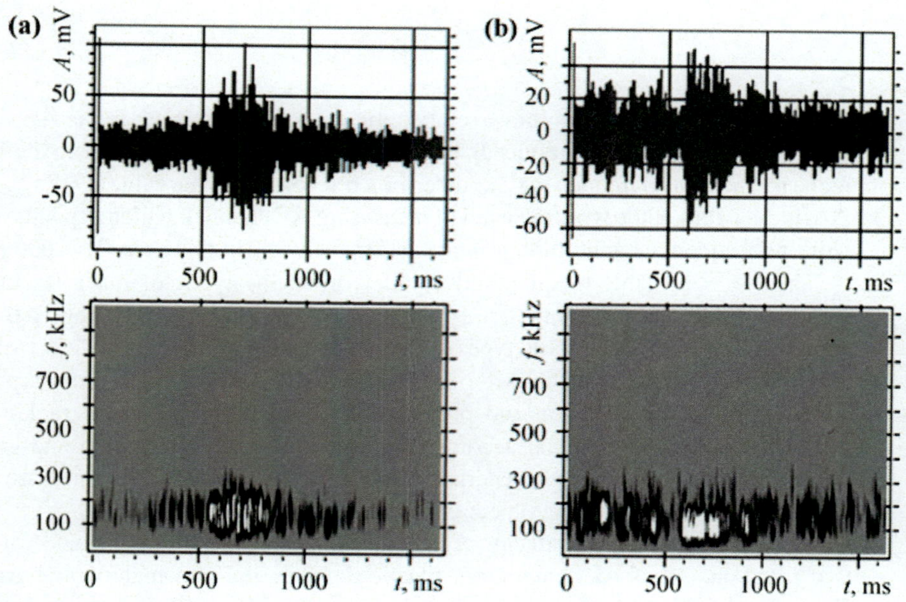

Fig. 1.12 Typical MAE signals and their wavelet transforms in the course of remagnetization parallel (**a**) and perpendicular (**b**) to the longitudinal direction of the axis of the strip

structures are detected in the early stages of their initiation what is important for diagnosing the state of such structural elements.

In [237], the possibility of using the MAE intensity for diagnostics of thick-walled objects is studied. It is found that FFT and wavelet analysis make it possible to separate the useful MAE signal, the intensity of which during diagnostics in industrial conditions may be lower than the background noise. A new filtering method provides measurement of only useful data, which greatly simplifies their analysis and increases the reliability of conclusions.

1.5.5 Application of the κ-Criterion for the Evaluation of the Fracture Types of Structural Materials

In the investigation of AE in the process of fracture of structural materials, it is important to be able to separate the origins of AE signals and reveal their correspondence to one or another type of fracture. Most of the known criteria for identifying the types of failure are based on the characteristics of the frequency spectrum of the signal obtained by the Fourier transform. In view of the high efficiency of WT in the up-to-date processing of AE signals, a new criterion of quantitative evaluation of the types of fracture is proposed in [238–240].

For this purpose, the researchers use the following criteria parameter κ:

$$\kappa = \frac{\mathrm{WT}_{\max} \times \Delta f_0}{\Delta f_{\max}}, \tag{1.9}$$

where $\mathrm{WT}_{\max}$ is the maximum wavelet coefficient of a local event in the AE signal, $\Delta f_{\max}$ is the width of the frequency band in the projection of the CWT on the WT–f plane for the central frequency $f_{\max}$, determined according to the value $0.7\mathrm{WT}_{\max}$; and Δf_0 is the width of the working band of measuring AE channel, which depends on the working frequency band of the primary AE transducer (in this case, 200–600 kHz).

Thus, it is possible to quantitatively determine the types of macrofracture of structural materials according to the parameter κ as follows: If $\kappa < 0.1$, then the AE signal characterizes the ductile type of fracture of the material (plastic deformation); if $0.1 \leq \kappa < 0.2$, then the AE signal characterizes the ductile–brittle type and corresponds to the initiation and propagation of microcracks (microcracking); if $\kappa \geq 0.2$, then the AE signal accompanies brittle fracture (formation and growth of macrocracks in structural materials). As the analyzed parameter increases, the probability of fracture of the object of monitoring increases.

Known methods of identifying AE signals are effective in the presence of only two mechanisms of AE generation: plastic deformation and macrocrack growth. However, in structural materials during their fracture, different processes can occur as AE sources: the formation of plastic deformations, microcracking, cracking of inclusions, macrofracture, and so on.

We have developed several methods for evaluation of the mechanisms of fracture of structural materials by wavelet analysis of AE signals. In particular, for their identification by AE signals based on CWT using a criteria parameter (1.9), it is necessary to construct CWT of AE signal and to determine the maximum value of the wavelet coefficient WT_{max} of the local event; to find a projection of the CWT on the plane WT–f at the moment of reaching the value of WT_{max}; determine the bandwidth Δf_{max} at the level $0.7WT_{max}$; to calculate the criteria index κ from relation (1.9) and by using its value to identify the type of fracture that occurs at a fixed time point in material [24, 176, 238–242].

The criterion for identifying the types of fracture by the spectral parameters of the AE signals is used during the diagnosis of oil pumping equipment [243]. AE signals with a criteria parameter within $0.2 \leq \kappa < 1.0$ are detected. According to the developed algorithm for estimating the types of fracture at such high values of the parameter, a conclusion is drawn that additional inspections of the dangerous area by other methods of NDT to identify and determine the size of crack-like defects are required. As a result of inspections by ultrasonic and magnetic methods, a propagating crack was found in the bend of the pipeline. To ensure further reliable operation of the facility, the area with the detected defect was replaced with a new element.

The proposed criterion was used to study the dynamics of the propagation of fatigue fracture in an aluminum alloy [244, 245]. A characteristic feature of AE at the stage of fatigue failure is that the signals are generated by certain groups that correspond to the formation of new surfaces near the crack front, and its area is proportional to the sum of the amplitudes of the registered signals. The transition from the initiation to stable propagation of fracture is characterized by a sharp jump in the amplitudes and number of AE signals.

The initiation and development of the fracture of dental polymers for temporary structures are studied with the help of the κ-criteria [241, 242]. It is found that under the action of load in polymers, different types of fracture occur (ductile, ductile–brittle, brittle), which alternate. According to the criteria parameter, the types of fracture of such materials under quasi-static tension are evaluated, and their ranking according to the brittleness index is carried out.

Since the constructed κ-criterion, as practice has shown, has an insignificant transition limit from the quantitative characteristics of ductile to brittle fracture, the purpose of further research is to develop an energy criterion and a new method of identifying the types of fracture of structural materials.

Thus, according to the analysis of literature, among the types of fracture of structural materials, the greatest attention is paid to the study of brittle (quasi-brittle) and ductile fracture. Brittle fracture is the most dangerous because it is spontaneous and very rapid. To improve the efficiency of TD and NDT of industrial facilities and products, it is important to be able to establish the type of fracture of the material. Today, the AE method is the most attractive one for studying the processes and stages of development of defects in the structure of materials and assessing the condition of real objects during their TD.

Combining analytical–numerical approaches to modeling and experimental to the sources identification of elastic waves AE, it is possible to establish relationships between the individual parameters of material defects and the characteristics of AE signals, which during TD and NDT can be used to solve the inverse problem—determination of the defect parameters by AE signals.

To analyze the structure and behavior of signals that contain areas of different duration with speed fluctuation or different frequency content, WT should be used. According to the literature, the use of AE and MAE signals is an effective tool of identifying the stages and mechanisms of fracture, assessing the degree of degradation of engineering materials which differ in nature and structure.

References

1. Cernak, E. A., & DelGrosso, E. J. (1960). *Brittle fracture: Ductile-brittle transactions, an annotated bibliography* (Vol. 301). PWAC, Pratt & Whitney Aircraft Division, United Aircraft Corporation, CANEL Operations.
2. González-Velázquez, J. L. (2018). *Fractography and failure analysis. Stuctral integrity* (Vol. 3). Springer International Publishing AG.
3. Liebowitz, H. (Ed.). (1969). *Fracture of metals*. Academic Press.
4. Thomason, P. F. (1990). *Ductile fracture of metals*. Pergamon Press.
5. Pocluda, J., & Šandera, P. (2010). *Micromechanisms of fracture and fatigue. In a multi-scale context*. Springer.
6. Pineau, A., Benzerga, A. A., & Pardoen, T. (2016). Failure of metals I: Brittle and ductile fracture. *Acta Materialia, 107*(1), 424–483.
7. Vladimirov, V. I. (1984). *Phizicheskaya priroda razrusheniya metallov (Physical nature of metal fracture)*. Metallurgiya.
8. Gdoutos, E. (1990). *Fracture mechanics criteria and applications. Engineering applications of fracture mechanics* (Vol. 10). Kluwer Academic Publishers.
9. Zehnder, A. T. (2012). *Fracture mechanics. Lectures notes in applied and computational mechanics* (Vol. 62). Springer.
10. Tang, H., Acharya, A., & Saigal, S. (2005). Directional dependence of crack growth along the interface of a bicrystal with symmetric tilt boundary in the presence of gradient effects. *Mechanics of Materials, 37*(5), 593–606.
11. Novikov, I. I. (1978). *Theory of heat treatment of metals*. Mir Publishers.
12. Meyers, M. A., & Chawla, K. K. (2009). *Mechanical behavior of materials*. Cambridge University Press.
13. McClintock, F. A., & Argon, A. S. (1966). *Mechanical behavior of materials*. Addison-Wesley Pub Co.
14. Rotshteyn, V. P. (2008). *Razrusheniye materialov i ispytaniya na udarnuyu vyazkost (Material fracture and impact tests)*. Publishing Center of TPU.
15. Drouillard, T. F. (1979). *Acoustic emission. A bibliography with abstracts*. IFI/Plenum.
16. Grosse, C. U., & Ohtsu, M. (2008). *Acoustic emission testing: Basics for research—Applications in civil engineering*. Springer.
17. Ono, K. (2011). Application of acoustic emission for structure diagnosis. *Diagnostics, 2*, 3–18.
18. Nazarchuk, Z., Skalskyi, V., & Serhiyenko, O. (2017). *Acoustic emission. Methodology and Application*. Springer International Publishing AG.
19. Skalskyi, V. R., & Koval, P. M. (2007). *Some methodological aspects of application of acoustic emission*. Publishing House Spolom.
20. Skalskyi, V. R., & Andreikiv, O. Y. (2006). *Otsinka obyemnoyi poshkodzhenosti materialiv metodom akustychnoyi emisiyi (Estimation of materials volume damage by acoustic emission method)*. Publishing Center LNU im. I. Franka.

21. Bozhydarnik, V. V., Skalskyi, V. R., & Matviiv, Yu. Y. (2012). *Diahnostuvannya ruinuvannya sklovolokonnych kompozytiv metodom akustychnoyi emisiyi (Diagnistics of fiber reinforced composites fracture by acoustic emission method)*. Naukova dumka.
22. Nazarchuk, Z. T., Andreikiv, O. Y., & Skalskyi, V. R. (2013). *Otsinyuvannya vodnevoyi dehradatsii feromahnetykiv u mahnetnomu poli (Estimation of hydrogen degradation of ferromagnets in a magnetic field)*. Naukova dumka.
23. Nazarchuk, Z. T., Nekludov, I. M., & Skalskyi, V. R. (2016). *Metod akustychnoyi emisiyi v diahnostuvanni korpusiv reaktoriv AES (Acoustic emission method in diagnosing NPP reactor buildings)*. Naukova dumka.
24. Skalskyi, V. R., Karpash, O. M., Koshovyi, V. V., Nedoseka, A. Y., & Stankevych, O. M. (2017). *Tekhnichna diahnostyka materialiv i konstrukcii, t 5: Akustychni metody kontrolyu dehradatsii materialiv i defektnosti elementiv konstrukcii (Technical diagnostics of materials and structures, vol. 5: Acoustic methods of control of materials degradation and structure elements damage.)*. Publishing House Prostir-M.
25. Nazarchuk, Z. T., Skalskyi, V. R., & Selivonchyk, T. V. (2019). *Akustyko-emisiine diagnostuvannya koroziino-vodnevogo roztriskuvannya konstrukciinych stalei (Acoustic emission diagnostics of corrosion-hydrogen cracking of structural steels)*. Naukova dumka.
26. Sharpe, J. A. (1942). The production of elastic waves by explosion pressures. I. Theory and empirical field observations. *Geophysics, 7*, 144–154.
27. Brock, L. M. (1981). The effects of displacement discontinuity derivatives on wave propagation. I: Three-dimensional elastic solid. *International Journal of Engineering Science, 19*(2), 243–252.
28. Brock, L. M. (1981). The effects of displacement discontinuity derivatives on wave propagation. II: Body force for a half-space. *International Journal of Engineering Science, 19*(6), 835–843.
29. Brock, L. M. (1981). The effects of displacement discontinuity derivatives on wave propagation. III: Body and point forces in elastic half-spaces. *International Journal of Engineering Science, 19*(7), 949–957.
30. Brock, L. M. (1982). The effects of displacement discontinuity derivatives on wave propagation. IV: Dislocation motion in a half-space. *International Journal of Engineering Science, 20*(3), 483–496.
31. Achenbach, J. D., & Harris, J. C. (1979). Acoustic emission from a brief crack propagation event. *Journal of Applied Mechanics, 46*(1), 107–112.
32. Scruby, C. B., Wadley, H. N. G., & Hill, J. J. (1983). Dynamic elastic displacement at the surface of an elastic half-space due to defect sources. *Journal of Physics D: Applied Physics, 16*, 1069–1083.
33. Danylenko, V. A., & Skurativskyi, S. I. (2016). Peculiarities of wave dynamics in media with oscillating inclusions. *International Journal of Non-Linear Mechanics, 84*, 31–38.
34. Sheikhhassani, R., & Dravinski, M. (2014). Scattering of a plane harmonic SH wave by multiple layered inclusions. *Wave Motion, 51*(3), 517–532.
35. Lee, J., Lee, H., & Jeong, H. (2016). Numerical analysis of SH wave field calculations for various types of a multilayered anisotropic inclusion. *Engineering Analusis with Boundary Elements, 64*, 38–67.
36. Ulitko, A. F., & Zrazhevskii, H. M. (1985). Izlucheniye uprugich voln pri skachkoobraznom dvizhenii polubeskonechnoho razreza (Emission of elastic waves in a jump-like motion of a semi-infinite cut). *Diagnostika i prognozirovaniye resursa cvarnych konstrukcii (Diagnostics and Prediction of the Service Life of Welded Structures), 1*, 23–24.
37. Flitman, L. V. (1983). Volny, vyzvannyye mgnovennym razryvom sploshnosti uprugoi sredy (Waves caused by instantaneous discontinuity of elastic medium). *Prikladnaya mechanika i matematika (Applied Mechanics and Mathematics), 27*(4), 618–628.
38. Alterman, Z., Burridge, R., & Loewentha, D. (1971). The vibration of elastic plane and half plane due to the sudden appearance of a crack. *Geophysical Journal of the Royal Astronomical Society, 24*(1/2), 239–259.

39. Saatov, Y. U., Bykovtsev, A. S., & Khamidov, L. A. (1985). *Seismologicheskiye zadachi mechaniki razrusheniya (Seismological problems of fracture mechanics)*. Mechmat.
40. Maslov, L. A. (1976). Model treshchiny kak izluchatelya uprugich kolebanii (Model of a crack as a radiator of elastic vibrations). *Prikladnaya matematika i tekhnicheskaya fizika (Applied Mathematics and Technicl Physics), 2*, 160–166.
41. Tripalin, A. S., & Builo, S. I. (1986). *Akusticheskaya emisiya: Fiziko-mekhanicheskiye aspekty (Acoustic emission: Physical and mechanical aspects)*. Publishing of Rostov University.
42. Jacobs, L. J., Scott, W. R., Granata, D. M., & Ryan, M. J. (1990). *Acoustic emission from a growing crack. Review of progress in quantitative nondestructive evaluation* (Vol. 9, pp. 1749–1756). Chimenti Plenum Press.
43. Maslov, L. A., & Apchipov, V. I. (1977). O diagramme napravlenosti izlucheniya ot treshchiny (About the radiation pattern from a crack). *Defektoskopiya (Defectoscopy), 5*, 26–29.
44. Andreykiv, O. Y., Lysak, M. V., Serhiyenko, O. M., & Skalsky, V. R. (2001). Analysis of acoustic emission caused by internal cracks. *Engineering Fracture Mechanics, 68*(11), 1317–1333.
45. Wadley, H. N. G., Simmons, J. A., & Turner, C. (1984). Predictive modeling of quantitative acoustic emission. Review of progress. In *Quantitative nondestructive evaluation* (Vol. 3B, pp. 683–697). Plenum Publishing Corporation.
46. Sinclair, J. E. (1979). Epicentre solution for point multipole sources in an elastic half-space. *Journal of Physics D: Applied Physics, 12*(8), 1309–1315.
47. Shibata, M. (1984). A theoretical evaluation of acoustic emission signals: The rise-time effect of dynamic forces. *Materials Evaluation, 42*(1), 107–120.
48. Gordeliy, E., & Detournay, E. (2011). Displacement discontinuity method for modeling axisymmetric cracks in an elastic half-space. *International Journal of Solids and Structures, 48*(19), 2614–2629.
49. Boström, A., Grahn, T., & Niklasson, A. J. (2003). Scattering of elastic waves by a rectangular crack in an anisotropic half-space. *Wave Motion, 38*(2), 91–107.
50. Mendelsohn, D. A., Achenbach, J. D., & Keer, L. M. (1980). Scattering of elastic waves by a surface-breaking crack. *Wave Motion, 2*(3), 277–292.
51. Pecorari, C. (2001). Scattering of a Rayleigh wave by a surface-breaking crack with faces in partial contact. *Wave Motion, 33*(3), 259–270.
52. Manolis, G. D., Dineva, P. S., & Rangelov, T. V. (2004). Wave scattering by cracks in inhomogeneous continua using BIEM. *International Journal of Solids and Structures, 41*(14), 3905–3927.
53. Zhang, C., & Achenbach, J. D. (1988). Scattering of body waves by an inclined surface-breaking crack. *Ultrasonics, 26*(3), 132–138.
54. Zhang, C., & Achenbach, J. D. (1988). Numerical analysis of surface-wave scattering by the boundary element method. *Wave Motion, 10*(4), 365–374.
55. Zeng, X., & Cakmak, A. S. (1984). Displacements and stresses caused by forces acting on the lateral surfaces of a crack in infinite elastic medium. *International Journal of Soil Dynamics and Earthquake Engineering, 3*(1), 42–48.
56. Iturrarán-Viveros, U., Sánchez-Sesma, F. J., & Luzón, F. (2008). Boundary element simulation of scattering of elastic waves by 3-D cracks. *Journal of Applied Geophysics, 64*(3/4), 70–82.
57. Andreykiv, O., Skalsky, V., Serhiyenko, O., & Rudavskyy, D. (2010). Acoustic emission estimation of crack formation in aluminium alloys. *Engineering Fracture Mechanics, 77*(5), 759–767.
58. Eskandari-Ghadi, M., Ardeshir-Behrestaghi, A., & Neya, B. N. (2013). Mathematical analysis for an axisymmetric disc-shaped crack in transversely isotropic half-space. *International Journal of Mechanical Sciences, 68*, 171–179.
59. Greshnikov, V. A., & Drobot, Yu. B. (1976). *Akusticheskaya emissiya (Acoustic emission)*. Publishing House of Standards.
60. Panasyuk, V. V. (Ed.). (2001). *Mechanika ruinuvannya i mitsnist materialiv, t. 5 (Fracture mechanics and strength of materials)* (Vol. 5). PhMI.

61. Vakar, K. B. (Ed.). (1980). *Akusticheskaya emissiya i eyo primineniye dlya nerazrushaushchego kontrolya v atomnoi energetike (Acoustic emission and its application for non-destructive testing in nuclear power)*. Atomizdat.
62. Borodachev, N. M. (2000). An elliptical crack under point forces applied to its surfaces. *Journal of Applied Mathematics and Mechanics, 64*(3), 477–483.
63. Zhong, X.-C., Li, X.-F., & Lee, K. Y. (2009). Analysis of a mode I crack perpendicular to an imperfect interface. *International Journal of Solids and Structures, 46*(6), 1456–1463.
64. Zhou, Z.-G., & Wu, L.-Z. (2007). Basic solutions of two parallel mode-I cracks or four parallel mode-I cracks in the piezoelectric materials. *Engineering Fracture Mechanics, 74*(9), 1413–1435.
65. Jin, B., & Zhong, Z. (2002). Dynamic stress intensity factor (mode I) of a penny-shaped crack in an infinite poroelastic solid. *International Journal of Engineering Science, 40*(6), 637–664.
66. Karapetian, E. N. (1993). The elastic field of an internal circular crack subjected to a linear normal loading. *Engineering Fracture Mechanics, 45*(5), 575–586.
67. Komvopoulos, K. (1996). Subsurface crack mechanisms under indentation loading. *Wear, 199*(1), 9–12.
68. Kim, Y. J., Kim, H.-G., & Im, S. (2001). Mode decomposition of three-dimensional mixed-mode cracks via two-state integrals. *International Journal of Solids and Structures, 38*(36–37), 6405–6426.
69. Wang, B. L., Han, J. C., & Du, S. Y. (2000). Fracture mechanics for multilayers with penny-shaped cracks subjected to dynamic torsional loading. *International Journal of Engineering Science, 38*(8), 893–901.
70. Lee, D.-S. (2009). Diffraction of torsional elastic waves by a peripheral edge crack around a spherical cavity. *International Journal of Solids and Structures, 46*(2), 433–439.
71. Harris, J. G., & Pott, J. (1984). Surface motion excited by acoustic emission from a buried crack. *Journal of Applied Mechanics, 51*(1), 77–83.
72. Shah, H., Wong, K. C., & Datta, S. K. (1985). Surface displacement due to elastic wave scattering by buried planar and non-planar cracks. *Wave Motion, 7*(4), 319–333.
73. Khai, M. V., & Stankevych, V. Z. (1994). Opredeleniye peremeshchenii na poverchnosti tela, obuslovlennych razryvami sploshnosti sredy (Determination of displacements on the surface of a body caused by discontinuities in the continuity of the medium). *Prikladnaya mechanika (Applied Mechanics), 30*, 25–33.
74. Khai, M. V., & Stankevych, V. Z. (2002). Fundamental solution of a quasistatic problem for a half space with two-dimensional discontinuities. *Materials Science, 38*(3), 338–343.
75. Mykhas'kiv, V. (2005). Transient response of plane rigid inclusion to an incident wave in an elastic solid. *Wave Motion, 41*, 133–144.
76. Stankevych, V. Z. (2008). Stress intensity nears a crack in the composition of a half space and a layer under harmonic loading. *Materials Science, 44*(2), 175–182.
77. Skalsky, V., Stankevych, O., & Serhiyenko, O. (2013). Wave displacement field at a half-space surface caused by an internal crack under twisting load. *Wave Motion, 50*(2), 326–333.
78. Stankevych, O., & Skalsky, V. (2017). The vibration of a half-space due to a buried mode I crack opening. *Wave Motion, 72*, 142–153.
79. Zhang, X., Feng, N., Wang, Y., & Shen, Y. (2014). An analysis of the simulated acoustic emission sources with different propagation distances, types and depths for rail defect detection. *Applied Acoustics, 86*, 80–88.
80. Cuadra, J., Vanniamparambil, P. A., Servansky, D., Bartoli, I., & Kontsos, A. (2015). Acoustic emission source modeling using a data-driven approach. *Journal of Sound and Vibration, 341*, 222–223.
81. Qui, T. B., & Rim, J.-M. (2021). Crack detection and localization in a fluid pipeline based on acoustic emission signals. *Mechanical System and Signal Processing, 150*, 107254.
82. Gerberich, W., Atteridge, D. G., & Lessar, J. F. (1975). An acoustic emission investigation of microscopic ductile fracture. *Metallurgical and Materials Transactions A: Physical Metallurgy and Materials Science, 6*(4), 797–801.

83. Yudin, A. A., & Ivanov, V. I. (1985). Acoustic emission in plastic deformation of metals (review). *Strength of Materials, 17*, 842–851.
84. Baram, J., & Rosen, M. (1981). Effect of grain size on the acoustic emission generated during plastic deformation of copper. *Materials Science and Engineering, 47*, 243–246.
85. Sikorski, W. (Ed.). (2013). *Acoustic emission: Research and application.* InTech.
86. Nakamura, Y., Veach, C. L., & McCauley, B. O. (1972). Amplitude distribution of acoustic emission signals. In R. Liptai, D. Harris, C. Tatro (Eds.), *Acoustic emission. ASTM International* (pp. 164–186). West Conshohocken, PA.
87. Mints, R. I., Kortov, V. S., & Melechin, V. P. (1973). Vliyaniye mechanizmov plasticheskoi deformacii na akusticheckiyu i ekzoelektronnuyu emissiyu (Influence of mechanisms of plastic deformation on acoustic and exoelectronic emission). In *Metallofizika (Metal physics)* (pp. 93–99). Naukova dumka.
88. Saklskyi, V. R., Botvina, L. R., & Stankevych, O. M. (2011). Diahnostuvannya mechanizmiv ruinuvannya stali 38HN3MFA za veivlet-peretvorennyam syhnaliv akustychnoi emisii (Diagnostics of fracture mechanisms of steel 38HN3MFA by wavelet transform of acoustic emission signals). *Tekhnicheskaya diagnostika i nerazrushayushchii kontrol (Technical Diagnostics and Non-destructive Testing), 3*, 12–17.
89. Scruby, C. B., Collingwood, J. C., & Wadley, H. N. G. (1978). A new technique for the measurement of acoustic emission transients and their relationship to crack propagation. *Journal of Physics D: Applied Physics, 11*, 2359–2369.
90. Brindley, B. J., Holt, J., & Palmer, I. G. (1973). Acoustic emission-3: The use of ringdown counting. *NDT, 6*(6), 299–306.
91. Kline, R. A., & Hartman, W. (1976). Frequency analysis of acoustic emission signals. In *Proceedings of the 2nd International conference on mechanical behaviour of materials*, Boston, MA, USA, 16–20 August 1976 (pp. 1631–1635).
92. Antonaci, P., Bocca, P., & Masera, D. (2012). Fatigue crack propagation monitoring by acoustic emission signal analysis. *Engineering Fracture Mechanics, 81*, 26–32.
93. Kuksenko, V. S., Lyashkov, A. I., & Savelyev, V. N. (1980). Akusticheskaya emissiya pri zarozhdenii i razvitii mikrotreshchin v stalyach (Acoustic emission during the initiation and development of microcracks in steels). *Defektoskopiya (Defectoscopy), 6*, 57–63.
94. Lyashkov, A. I., Inzhevatkin, I. E., & Savelyev, V. N. (1980). Izucheniye zarozhdeniya mikrotreshchin v metallach metodom akusticheskoi emissii (Study of the initiation of microcracks in metals by the acoustic emission method). *Defektoskopiya (Defectoscopy), 6*, 98–101.
95. Woodward, B. (1976). Identification of acoustic emission source mechanisms by energy spectrum analysis. *Ultrasonics, 14*(6), 249–255.
96. Fleischman, P., Rouby, D., Lakestani, F., & Baboux, J. C. (1975). A spectrum analysis of acoustic emission. *NDT, 8*, 241–244.
97. Rouby, D., & Fleischmann, P. (1978). Spectral analysis of acoustic emission from aluminium single crystals undergoing plastic deformation. *Physica Status Solidi A, 48*(2), 439–445.
98. Raj, B., Jha, B. B., & Rodriguez, P. (1989). Frequency spectrum analysis of acoustic emission signal obtained during tensile deformation and fracture of an AISI 316 type stainless steel. *Acta Metallurgica, 37*(8), 2211–2215.
99. Bunina, N. A. (1990). *Issledovaniye plasticheskoi deformacii metallov metodom akusticheskoi emissii (Study of the plastic deformation of metals by the acoustic emission method).* Publishing Center of Leningrad University.
100. Takahashi, H., Khan, M. A., Kikuchi, M., & Suzuki, M. (1981). Acoustic emission crack monitoring in fracture-toughness tests for AISI-4340 and SA 533B steels. *Experimental Mechanics, 21*(3), 89–99.
101. Muravin, G. B., Finkel, V. M., Lezvinskaya, L. M., & Simkin, Y. V. (1984). Issledovaniye deformirovaniya kremnistoho zheleza metodom akusticheskoi emissii (Investigation of the deformation of siliceous iron by the acoustic emission method). *Defektoskopiya (Defectoscopy), 10*, 88–91.

102. Braginskii, A. P. (1984). Raspoznavaniye defektov po spektralnym charakteristikam akusticheskoi emissii (Defect recognition by spectral characteristics of acoustic emission). *Defektoskopiya (Defectoscopy), 1*, 47–55.
103. Pardee, W., & Graham, L. J. (1975), Analysis of acoustic emission frequency spectra. In *IEEE Ultrasonic symposium proceedings* (pp. 597–600). Los Angeles.
104. Ivanov, V. I. (1980). Primeneniye metoda AE dlya nerazrushayushchego kontrolya i issledovaniya materialov (obzor osnovnykh problem i zadach) (Application of the AE method for non-destructive testing and research of materials (review of the main problems and tasks)). *Defektoskopiya (Defectoscopy), 5*, 65–84.
105. Vanniamparambil, P. A., Guclu, U., & Kontsos, A. (2015). Identification of crack initiation in aluminum alloys using acoustic emission. *Experimental Mechanics, 55*(5), 837–850.
106. Kostryzhev, A. G., Punch, R. B., Davis, C. L., & Strangwood, M. (2012). Acoustic emission monitoring of split formation during charpy impact testing of high strength steel. *Materials Science and Technology, 28*(2), 240–242.
107. Lysak, M. V. (1994). Acoustic emission during jumps in subcritical growth of crack in three-dimensional bodies. *Engineering Fracture Mechanics, 47*(6), 873–879.
108. Friesel, M. A. (1996). Acoustic emission source identification using longwaveguide sensors. *NDT & International, 19*(3), 203–206.
109. Vinogradov, A., Orlov, D., Danyuk, A., & Estrin, Y. (2013). Effect of grain size on the mechanisms of plastic deformation in wrought Mg–Zn–Zr alloy revealed by acoustic emission measurements. *Acta Materialia, 61*(6), 2044–2056.
110. Inoue, H., Tsuzuki, R., Nohara, S., & Iwakura, C. (2007). Characterization of initial activation behavior for hydrogen storage alloys by acoustic emission technique. *Journal of Alloys and Compounds, 446–447*, 681–686.
111. Hase, A., Mishina, H., & Wada, M. (2012). Correlation between features of acoustic emission signals and mechanical wear mechanisms. *Wear, 292–293*, 144–150.
112. Meriaux, J., Boinet, M., Fouvry, S., & Lenain, J. C. (2010). Identification of fretting fatigue crack propagation mechanisms using acoustic emission. *Tribology International, 43*(11), 2166–2174.
113. Amer, A. O., Gloanec, A.-L., Courtin, S., & Touze, C. (2013). Characterization of fatigue damage in 304L steel by an acoustic emission method. *Procedia Engineering, 66*, 651–660.
114. Friedrich, K. (Ed.). (1989). Application of fracture mechanics to composite materials. In *Composite materials series* (Vol. 6). Elsevier Science Publishers.
115. Guagliano, M., & Aliabadi, M. H. (Eds.). (2005). *Fracture and damage of composites*. WIT Press.
116. Talreja, R., & Singh, C. V. (2012). *Damage and failure of composite materials*. Cambridge University Press.
117. De Rosa, I. M., Santulli, C., & Sarasini, F. (2009). Acoustic emission for monitoring the mechanical behaviour of natural fibre composites: A literature review. *Composites Part A: Applied Science and Manufacturing, 40*, 1456–1469.
118. Lomov, S. V., Bogdanovich, A. E., Ivanov, D. S., Mungalov, D., Karahan, M., & Verpoest, I. (2009). A comparative study of tensile properties of non-crimp 3D orthogonal weave and multi-layer plain weave E-glass composites. Part 1: Materials, methods and principal results. *Composites Part A: Applied Science and Manufacturing, 40*, 1134–1143.
119. Ivanov, D. S., Lomov, S. V., Bogdanovich, A. E., Karahan, M., & Verpoest, I. (2009). A comparative study of tensile properties of non-crimp 3D orthogonal weave and multi-layer plain weave E-glass composites. Part 2: Comprehensive experimental results. *Composites Part A: Applied Science and Manufacturing, 40*, 1144–1157.
120. De Rosa, I. M., Santulli, C., Sarasini, F., & Valente, M. (2009). Post-impact damage characterization of hybrid configurations of jute/glass polyester laminates using acoustic emission and IR thermography. *Composites Science and Technology, 69*, 1142–1150.
121. Ivanov, D. S., Baudri, F., Van Den Broucke, B., Lomov, S. V., Hie, H., & Verpoest, I. (2009). Failure analysis of triaxial braided composite. *Composites Science and Technology, 69*, 1372–1380.

122. Truong, T. C., Inavov, D. S., Klimshin, D. V., & Verpoest, I. (2008). Carbon composites based on multi-axial multi-ply stitched performs. Part 7: Mechanical properties and damage observations in composites with sheared reinforcement. *Composites Part A: Applied Science and Manufacturing, 39*, 1380–1393.
123. Böhm, R., & Hufenbach, W. (2010). Experimentally based strategy for damage analysis of textile-reinforced composites under static loading. *Composites Science and Technology, 70*, 1330–1337.
124. Bakhtiary Davijani, A. A., Hajikhani, M., & Ahmadi, M. (2011). Acoustic emission based on sentry function to monitor the initiation of delamination in composite materials. *Materials & Design, 32*, 3059–3065.
125. Oskouei, A. R., Zucchelli, A., Ahmadi, M., & Minak, G. (2011). An integrated approach based on acoustic emission and mechanical information to evaluate the delamination fracture toughness at mode I in composite laminate. *Materials & Design, 32*, 1444–1455.
126. Aggelis, D. G., Soulioti, D. V., Sapouridis, N., Burkola, N. M., Paipetis, A. S., & Matikas, T. E. (2011). Acoustic emission characterization of the fracture process in fibre reinforced concrete. *Construction and Building Materials, 25*, 4126–4131.
127. Mechraoui, S.-E., Laksimi, A., & Benmedakhene, S. (2012). Reliability of damage mechanism localisation by acoustic emission on glass/epoxy composite material plate. *Composite Structures, 94*, 1483–1494.
128. Sause, M. G. R., Müller, T., Horoschenkoff, A., & Horn, S. (2012). Quantification of failure mechanisms in mode-I loading of fiber reinforced plastics utilizing acoustic emission analysis. *Composites Science and Technology, 72*, 167–1741.
129. Liu, P. F., Chu, J. K., Lu, Y. L., & Zheng, J. Y. (2012). A study on the failure mechanisms of carbon fiber/epoxy composite laminates using acoustic emission. *Materials & Design, 37*, 228–235.
130. Njuhovic, E., Bräu, M., Wolff-Fabris, F., Starzynski, K., & Altstädt, V. (2015). Identification of failure mechanisms of metallised glass fibre reinforced composites under tensile loading using acoustic emission analysis. *Composites Part B Engineering, 81*, 1–13.
131. Crivelli, D., Guagliano, M., Eaton, M., Pearson, M., Al-Jumaili, S., Holford, K., & Pullin, R. (2015). Localisation and identification of fatigue matrix cracking and delamination in a carbon fibre panel by acoustic emission. *Composites Part B Engineering, 74*, 1–12.
132. Farnam, Y., Geiker, M. R., Bentz, D., & Weiss, J. (2015). Acoustic emission waveform characterization of crack origin and mode in fractured and ASR damaged concrete. *Cement and Concrete Composites, 60*, 135–145.
133. Woo, S.-C., & Goo, N.-S. (2007). Analysis of the bending fracture process for piezoelectric composite actuators using dominant frequency bands by acoustic emission. *Composites Science and Technology, 67*(7–8), 1499–1508.
134. Woo, S.-C., & Kim, T.-W. (2014). High-strain-rate impact in Kevlar-woven composites and fracture analysis using acoustic emission. *Composites Part B Engineering, 60*, 125–136.
135. Woo, S.-C., & Choi, N.-S. (2007). Analysis of fracture process in single-edge-notched laminated composites based on the high amplitude acoustic emission events. *Composites Science and Technology, 67*, 1451–1458.
136. Bussiba, A., Kupiec, M., Ifergane, S., Piat, R., & Böhlke, T. (2008). Damage evolution and fracture events in various composites by acoustic emission technique. *Composites Science and Technology, 68*, 1144–1155.
137. Fang, P., Cheng, L., Zhang, L., & Nie, J. (2008). Monotonic tensile behavior analysis of three-dimensional needle-punched woven C/SiC composites by acoustic emission. *Journal of University of Science and Technology Bejijng, 15*(3), 302–306.
138. Sause, M. G. R., Haider, F., & Horn, S. (2009). Quantification of metallic coating failure on carbon fiber reinforced plastics using acoustic emission. *Surface and Coating Technology, 204*, 300–308.
139. Gutkin, R., Green, C. J., Vangrattanachai, S., Pinho, S. T., Robinson, P., & Curtis, P. T. (2011). On acoustic emission for failure investigation in CFRP: Pattern recognition and peak frequency analyses. *Mechanical System and Signal Processing, 25*, 1393–1407.

140. De Groot, P. J., Wijnen, P. A. M., & Janssen, R. B. F. (1995). Real-time frequency determination of acoustic emission for different fracture mechanisms in carbon/epoxy composites. *Composites Science and Technology, 55*, 405–412.
141. Ramirez-Jimenez, C. R., Papadakis, N., Reynolds, N., Gan, T. H., Purnell, P., & Pharaoh, M. (2004). Identification of failure modes in glass/polypropylene composites by means of the primary frequency content of the acoustic emission even. *Composites Science and Technology, 64*, 1819–1827.
142. Shibata, M., & Ono, K. (1981). Magnetomechanical acoustic emission—A new method of nondestructive stress measurement. *NDT & International, 14*(5), 227–234.
143. Jiles, D. C. (1988). Review of magnetic methods for nondestructive evaluation. *NDT & International, 21*(5), 311–319.
144. Ng, D. H. L., Jacubovitc, J. P., Scruby, C. B., & Briggs, G. A. D. (1992). Effect of stress on magnetoacoustic emission from mild steel and nickel. *Journal of Magnetism and Magnetic Materials, 104*, 355–356.
145. Khamitov, V. A., Gorkunov, E. S., & Bartenev, O. A. (1988). Effect of elastic stress on magnetoelastic acoustic emission. *Defektoskopiya, 9*, 3–10.
146. Su, F. (2014). Methodology for the stress measurement of ferromagnetic materials by using magneto-acoustic emission. *Experimental Mechanics, 54*(8), 1431–1439.
147. Dhar, A., Jagadish, C., & Atherton, D. L. (1991). The effect of sample size on magnetoacoustic emission. *NDT & Intetnational, 24*(8), 15–19.
148. O'Sullivan, D., Cotterell, M., Tanner, D. A., Tanner, D. A., & Mészáros, I. (2004). Magneto-acoustic emission for the characterization of ferritic stainless steel microstructural state. *Journal of Magnetism and Magnetic Materials, 271*, 381–389.
149. Piotrowski, L., Augustyniak, B., Chmielewski, M., & Łabanowski, J. (2012). Study on the applicability of the measurements of magnetoelastic properties for a nondestructive evaluation of thermally induced microstructure changes in the P91 grade steel. *NDT E International, 47*, 157–162.
150. Kim, H. C., & Kim, C. G. (1989). Effect on magnetization frequency and stress on magnetoacoustic emission from 3% Si–Fe crystals. *Journal of Physics D Applied Physics, 22*(1), 192–198.
151. Dhar, A., & Atherton, D. L. (1992). Magnetizing frequency dependence of magnetoacoustic emission in pipeline steel. *IEEE Transactions on Magnetics, 28*(2), 1003–1007.
152. Ng, D. H. L., Yu, C. C., Qin, C.-D., Lo, C. C. H., & Jakubovics, J. P. (1996). The dependence of magnetoacoustic emission on magnetizing frequency in nickel and mild steel. *Journal of Applied Physics, 79*(8), 6057–6059.
153. Wilson, J. W., Tian, G. Y., Vaidhianathasamy, M., & Shaw, B. A. (2009). Magneto-acoustic emission and magnetic Barkhausen emission for case depth measurement in En36 gear steel. *IEEE Transactions on Magnetics, 45*(1), 177–183.
154. Perevertov, O., & Stupakov, A. (2015). Magnetoacoustic measurements on steel samples at low magnetizing frequencies. *Journal of Electrical Engineering, 66*(7/s), 58–61.
155. Kameda, J. (1988). Characterization of tempered martensite microstructure and embrittlement by acoustic emission and magnetic Barkhausen signal measurement. *Scripta Metallurgica, 22*(9), 1487–1492.
156. Piotrowski, L., Chmielewski, M., & Augustyniak, B. (2012). The influence of elastic deformation on the properties of the magnetoacoustic emission (MAE) signal for GO electrical steel. *Journal of Magnetism and Magnetic Materials, 324*(16), 2496–2500.
157. Gorkunov, E. S., Dragoshanskii, Yu. N., & Bartenev, O. A. (2001). Magnetoelastic acoustic emission in ferromagnetic materials. II. Effect of elastic and plastic strains on parameters of magnetoeastic acoustic emission. *Russian Journal of Nondectructive Testing, 37*, 835–858.
158. Ng, D. H. L., Lo, C. C., & Jakubovics, J. P. (1994). The effect of demagnetizing and stray fields on magnetoacoustic emission. *Journal of Applied Physics, 75*(10), 7009–7011.
159. Jadabach, C., Clapham, L., & Atherton, D. L. (1990). Orientation effect of anisotropy, stress, excitation, bias, and residual field on Barkhausen noise generation in pipeline steel. *Journal of Physics D Applied Physics, 23*(4), 443–447.

160. Park, D. G., Ok, C. I., Jeong, H. T., Kuk, I. H., & Hong, J. H. (1999). Nondestructive evaluation of irradiation effects in RPV steel using Barkhausen noise and magnetoacoustic emission signals. *Journal of Magnetism and Magnetic Materials, 196–197*, 382–384.
161. Edwards, C., & Palmer, S. B. (1987). Effect of stress and sample shape on the magnitude and frequency of magnetomechanical acoustic emission. *Journal of the Acoustical Society of America, 82*(2), 534–544.
162. Augustyniak, B., Piotrowski, L., & Chmielewski, M. (2008). Impact of frequency and sample geometry on magnetoacoustic voltage properties for two steel grades. *Journal of Electrical Engineering, 59*(7/s), 33–36.
163. Ivanchenko, S. V., Grokhovsky, V. I., & Kolchanov, N. N. (2016). Frequency spectra of magnetoacoustic emission in meteorites. In *79th annual meeting of the meteoritical society* (p. 6445), Berlin, 7–12 August 2016.
164. Kostin, V. N., Guryev, M. A., Vasilenko, O. N., et al. (2013). Amplitudno-chastotnyie charakteristiki magnitoakesticheskoi emissii termoobrabotannych splavov zheleza (Amplitude-frequency characteristics of magnetoacoustic emission of heat-treated iron alloys). *Fizicheskaya mekhanika (Physical Mechanics), 16*(5), 103–110.
165. Namkung, M., Wincheski, B., Fulton, J. P., & DeNale, R. (1992). AC magnetic field frequency dependence of magnetoacoustic emission. In *IEEE ultrasonics symposium proceedings*, 20–23 Oct. 1992, Tucson, AZ, USA. Retrieved 23 March, 2021, from 20040129648.pdf (nasa.gov)
166. Fourier, J. (1878). *The analytical theory of heat.* Cambridge University Press.
167. Daubechies, I. (1992). *Ten lectures on wavelets.* SIAM.
168. Mallat, S. G. (1998). *A wavelet tour of signal processing.* Academic Press.
169. Chui, C. K. (1992). *An introduction to wavelets.* Academic Press.
170. Blatter, K. (1998). *Wavelets: A primer.* AK Peters/CRC Press.
171. Hubbard, B. B. (1998). *The world according to wavelets: The story of a mathe matical technique in the making.* AK Peters/CRC Press.
172. Davydov, A. V. (2005). *Veivletnyye preobrazovaniya signalov (Wavelet transforms of signals).* UGGU, IGiG.
173. Wang, Q., & Chu, F. (2001). Experimental determination of the rubbing location by means of acoustic emission and wavelet transform. *Journal of Sound and Vibration, 248*(1), 91–103.
174. Li, Y., & Yi-Chu, Z. (2006). Wavelet analysis of acoustic emission signals from thermal barrier coatings. *Transactions of Nonferrous Metals Society of China, 16*, 270–275.
175. Tang, Y. Y., Liu, J., Yahg, L. H., & Ma, H. (2000). *Wavelet theory and its applications to pattern recognition.* World Scientific.
176. Skalsky, V., Stankevych, O., & Kuz, I. S. (2018). Application of wavelet transforms for the analysis of acoustic-emission signals accompanying fracture processes in materials (A survey). *Materials Science, 54*(2), 139–153.
177. Astaf'eva, N. M. (1996). Wavelet analysis: Basic theory and some applications. *Physics-Uspechi, 39*(11), 1085–1108.
178. Babicheva, I. F., & Sharko, A. V. (2005). Teoreticheskiye razrabotki po ispolzovaniyu veivlet-analiza i neirosetevych technologii v sisteme diagnostiki i prognozirovaniya ostatochnogo resursa promyshlennogo oborudovaniya (Theoretical developments on the use of wavelet analysis and neural network technologies in the system of diagnostics and forecasting of the residual life of industrial equipment). *Tekhnicheskaya diagnostika i nerazrushayushchii kontrol (Technical Diagnostics and Non-destructive Testing), 2*, 17–21.
179. Kunpeng, Z., San, W. Y., & Soon, H. G. (2009). Wavelet analysis of sensor signals for tool condition monitoring: A review and some new results. *Internatinal Journal of Machine Tools & Manufacturing, 49*, 537–553.
180. Li, X. (2002). A brief review: Acoustic emission method for tool wear monitoring during turning. *Internatinal Journal of Machine Tools & Manufacturing, 42*, 157–165.
181. Peng, Z. K., & Chu, F. L. (2004). Application of the wavelet transform in machine condition monitoring and fault diagnostics: A review with bibliography. *Mechanical System and Signal Processing, 18*, 199–221.

182. Bianchi, D., Mayrhofer, E., Gröschl, M., Betz, G., & Vernes, A. (2015). Wavelet packet transform for detection of single events in acoustic emission signals. *Mechanical System and Signal Processing, 64*, 441–451.
183. Saedi, F., Shevchik, S. A., & Wasmer, K. (2016). Automatic detection of scuffing using acoustic emission. *Tribology International, 94*, 112–117.
184. Zhang, X., Feng, N., Wang, Y., & Shen, Y. (2015). Acoustic emission detection of rail defect based wavelet transform and Shannon entropy. *Journal of Sound and Vibration, 339*, 419–432.
185. Hemmati, F., Orfali, W., & Gadala, M. S. (2016). Roller bearing acoustic signature extraction by wavelet packet transform, applications in fault detection and size estimation. *Applied Acoustic, 104*, 101–118.
186. Skrickij, V., Bogdevičius, M., & Junevičius, R. (2016). Diagnostic features for the condition monitoring of hypoid gear utilizing the wavelet transform. *Applied Acoustic, 106*, 51–62.
187. Dias, E. A., Pereira, F. B., Filho, S. L. M. R., & Brandaõ, L. C. (2016). Monitoring of through-feed centreless grinding processes with acoustic emission signals. *Measurement, 94*, 71–79.
188. Morizet, N., Godin, N., Tang, J., & Normand, B. (2016). Classification of acoustic emission signals using wavelets and random forests: Application to localized corrosion. *Mechanical System and Signal Processing, 70–71*, 1026–1037.
189. Wang, X. H., Zhu, C. M., Mao, H. L., & Huang, Z. F. (2009). Wavelet packet analysis for the propagation of acoustic emission signals across turbine runners. *NDT E International, 42*, 42–46.
190. Stepanova, L. N., Ramazanov, I. S., Kabanov, S. I., et al. (2007). Ispolzovaniye veivlet-filtracii pri loaklozacii signalov akusticheskoi emissii (Using wavelet filtering for localization of acoustic emission signals). *Kontrol. Diagnostika (Contrlol. Diagnostics), 9*, 27–31.
191. Stepanova, L. N., Ramazanov, I. S., & Kabanov, S. I. (2008). Veivlet-filtrasiya v zadachach lokalizacii signalov akusticheskoi emissii (Wavelet filtering in problems of localization of acoustic emission signals). *Kontrol. Diagnostika (Contrlol. Diagnostics), 1*, 15–19.
192. Douka, E., Loutridis, S., & Trochidis, A. (2004). Crack identification in plates using wavelet analysis. *Journal of Sound and Vibration, 270*, 279–295.
193. Terentyev, D. A., & Elizarov, S. V. (2008). Veivlet-analiz signalov AE v tonkostennych obyektach (Wavelet analysis of AE signals in thin-walled objects). *Kontrol. Diagnostika (Contrlol. Diagnostics), 7*, 51–54.
194. Jiao, J., He, C., Wu, B., Fei, R., & Wang, X. (2004). Application of wavelet transform on modal acoustic emission source location in thin plates with one sensor. *International Journal of Pressure Vessels and Piping, 8*, 427–431.
195. Gaul, L., & Hurlebaus, S. (1998). Identification of the impact location on a plate using wavelets. *Mechanical System and Signal Processing, 12*(6), 783–795.
196. Hamstad, M. A., O'Gallagher, A., & Gary, J. (2002). A wavelet transform applied to acoustic emission signals: Part 2: Source location. *Journal of Acoustic Emission, 20*, 62–82.
197. Zhang, Y., Lu, W., & Chu, F. (2017). Planet gear fault localization for wind turbine gearbox using acoustic emission signals. *Renewable Energy, 109*, 449–460.
198. Pomponi, E., Vingradov, A., & Danyuk, A. (2015). Wavelet based approach to signal activity detection and phase picking: Application to acoustic emission. *Signal Processing, 115*, 110–119.
199. Liu, X., Xiao, D., Shan, Y., Pan, Q., He, T., & Gao, Y. (2017). Solder joint failure localization of welded joint based on acoustic emission beamforming. *Ultrasonic, 74*, 221–232.
200. Sagasta, F., Zitto, M. E., Piotrowski, M., Benavent-Climent, A., Suarez, E., & Galeggo, A. (2018). Acoustic emission energy b-value for local damage evaluation in reinforced concrete structures subjected to seismic loadings. *Mechanical System and Signal Processing, 102*, 262–277.
201. Hamstad, M. A., O'Gallagher, A., & Gary, J. (2002). A wavelet transform applied to acoustic emission signals: Part 1: Source identification. *Journal of Acoustic Emission, 20*, 39–61.
202. Khamedi, R., Fallahi, A., & Oskouei, A. R. (2010). Effect of martensite phase volume fraction on acoustic emission signals using wavelet packet analysis during tensile loading of dual phase steels. *Materials & Design, 31*, 2752–2759.

203. Grabowska, J., Palacz, M., & Krawczuk, M. (2008). Damage identification by wavelet analysis. *Mechanical System and Signal Processing, 22*, 1623–1635.
204. Kim, H., & Melhem, H. (2004). Damage detection of structures by wavelet analysis. *Engineering Structures, 26*, 347–362.
205. Piotrkowski, R., Castro, E., & Gallego, A. (2009). Wavelet power, entropy and bispectrum applied to AE signals for damage identification and evaluation of corroded galvanized steel. *Mechanical System and Signal Processing, 23*, 432–445.
206. Ahadi, M., & Bakhtiar, M. S. (2010). Leak detection in water-filled plastic pipes through the application of tuned wavelet transforms to acoustic emission signals. *Applied Acoustics, 71*, 634–639.
207. Chen, C., Kovacevic, R., & Jandgric, D. (2003). Wavelet transform analysis of acoustic emission in monitoring friction stir welding of 6061 aluminium. *International Journal of Machine Tools & Manufacturing, 43*, 1383-1390.
208. Piotrkowski, R., Callego, A., Castro, E., García-Hernandez, M. T., & Ruzzante, J. E. (2005). Ti and Cr nitride coating/steel adherence assessed by acoustic emission wavelet analysis. *NDT E International, 38*, 260–267.
209. Gallego, A., Gil, J. F., Vico, J. M., Ruzzante, Y. E., & Piotrowski, R. (2005). Coating adherence in galvanized steel assessed by acoustic emission wavelet analysis. *Scripta Materialia, 52*, 1069–1074.
210. Feng, Y., & Schlindwein, F. S. (2009). Normalized wavelet packets quantifiers for condition monitoring. *Mechanical System and Signal Processing, 23*, 712–723.
211. Liu, Q., Chen, X., & Gindy, N. (2006). Investigation of acoustic emission signals under a simulative environment of grinding burn. *International Journal of Machine Tools & Manufacturing, 46*, 284–292.
212. Loutridis, S., Douka, E., Hadjileontiadis, L. J., & Trochidis, A. (2005). A two-dimensional wavelet transform for detection of cracks in plates. *Engineering Structures, 27*, 1327–1338.
213. Ni, Q.-Q., & Iwamoto, M. (2002). Wavelet transform of acoustic emission signals in failure of model composites. *Engineering Fracture Mechanics, 69*, 717–728.
214. Jonson, M. (2003). Classification of AE transients based on numerical simulations of composite laminates. *NDT E International, 36*, 319–329.
215. Qi, G., Barhorst, A., Hashemi, J., & Kamala, G. (1997). Discrete wavelet decomposition of acoustic emission signals from carbon-fiber-reinforced composite. *Composites Science and Technology, 57*, 389–403.
216. Qi, G. (2000). Wavelet-based AE characterization of composite materials. *NDT E International, 33*, 133–144.
217. Jeong, H., & Jang, Y. S. (2000). Wavelet analysis of plate wave propagation in composite laminates. *Composite Structures, 49*, 443–450.
218. Jeong, H. (2001). Analysis of plate wave propagation in anisotropic laminates using a wavelet transform. *NDT E International, 34*, 185–190.
219. Ding, Y., Reuben, R. L., & Steel, J. A. (2004). A new method for waveform analysis for estimating AE wave arrival times using wavelet decomposition. *NDT E International, 37*, 279–290.
220. Ciampa, F., & Meo, M. (2010). A new algorithm for acoustic emission localization and flexural group velocity determination in anisotropic structures. *Composites Part A, 41*, 1777–1786.
221. Kalogiannakis, G., Quintelier, J., De Baets, P., Degrieck, J., & Van Hemelrijck, D. (2008). Identification of wear mechanisms of glass/polyester composites by means of acoustic emission. *Wear, 264*, 235–244.
222. Sung, D.-U., Kim, C.-G., & Hong, C.-S. (2002). Monitoring of impact damages in composite laminates using wavelet transform. *Composites Part B Engineering, 33*, 35–43.
223. Marec, A., Thomas, J.-H., & El Guerjouma, R. (2008). Damage characterization of polymer-based composite materials: Multivariable analysis and wavelet transform for clustering acoustic emission data. *Mechanical System and Signal Processing, 22*, 1441–1464.
224. Loutas, T. H., Kostopoulos, V., Ramirez-Jimenez, C., & Pharaoh, M. (2006). Damage evolution in center-holed glass/polyester composites under quasi-static loading using

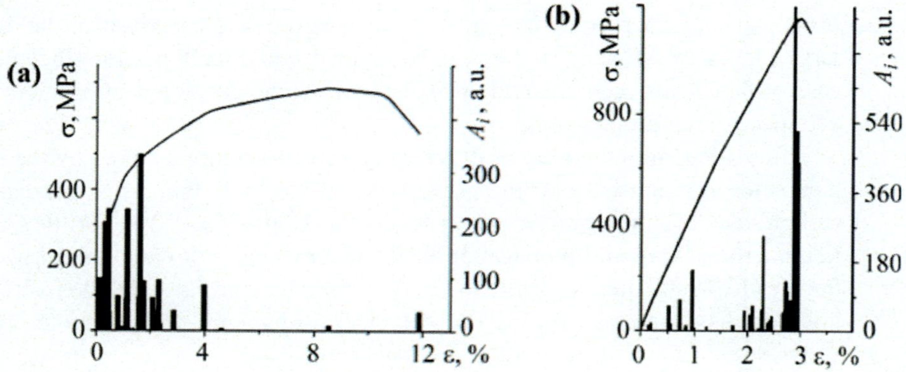

Fig. 3.7 Typical stress–strain curve and distribution of the AE signal amplitudes for carbon steel (**a**) and hardened carbon steel (**b**)

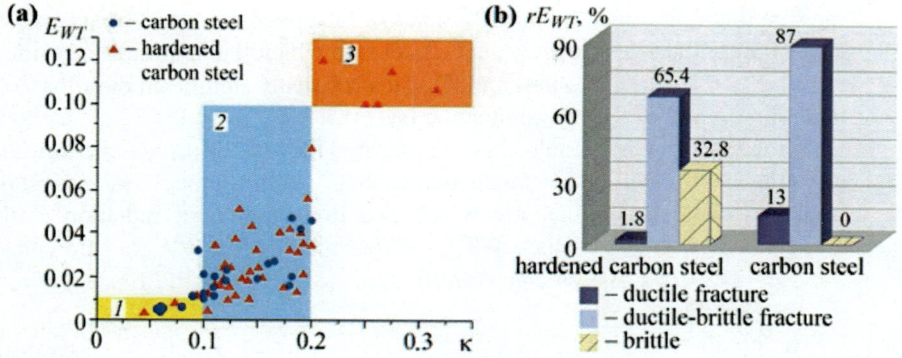

Fig. 3.8 Two-parameter distribution of AE signals during the fracture of two type carbon steel (**a**) (1—ductile fracture, 2—ductile–brittle fracture, 3—brittle fracture) and the energy ratio of the different types of fracture (**b**)

$E_{WT} \leq 0.0098$ for ductile, $0.01 \leq E_{WT} \leq 0.08$ for ductile–brittle and $0.1 \leq E_{WT} \leq 0.12$ for brittle.

It is known [21] that the signal from the crack can be preceded by signals from dislocations emitted by the plastically deformed area at its tip. The pulse duration of continuous AE is small (<4 μs), so such AE can be associated with the propagation of dislocations [41]. The AE, caused by the movement of dislocations, quickly disappears in a solid, thus indicating a high frequency of such signals [16]. The appearance of the second peak in the acoustic signal spectrum at a much lower frequency than the main one means the transition of crack formation from the stage of diffusion accumulation of defects to their enlargement and clustering [42]. Considering the above said, ductile fracture is understood as the processes associated with active plastic deformation (sliding or twinning of dislocations), formation, and growth of pores. Ductile–brittle fracture is considered as the initiation and growth of microcracks, and brittle—as the macrocrack propagation. Thus, based on the analysis of the parameters of CWT of AE signals, they can be divided into three types for efficiency and ease of diagnosis: accompanying plastic deformation, development of microcracks and macrocrack propagation.

After calculating the ratio of different types of fracture identified by the energy parameter and constructing the curves (Fig. 3.8b), it is found that for hardened carbon steel, the major fracture types are ductile–brittle (65.4%) and brittle (32.8%), i.e., microcracking and propagation of the macrocrack. For carbon steel, we have ductile (13%) and ductile–brittle (87%), i.e., the mechanisms of ductile fracture and microcracks formation prevail. The obtained results agree well with the obtained stress–strain curves of the studied materials (Fig. 3.7).

Figure 3.9 shows the CWT of typical AE signals for brittle and ductile fracture. The specific feature of the CWT of AE signals that accompany the brittle fracture (Fig. 3.9a) is a relatively large value of the wavelet coefficient WT_{max}, the narrow frequency band Δf in the WT–f projection at the level 0.7 WT_{max} (the bandwidth of the signal is determined by the difference of the upper and lower frequencies of the area of amplitude–frequency characteristics, at which the amplitude of oscillations is $1/\sqrt{2} \approx 0.7$ of the maximum [43]) and a relatively significant duration Δt of its radiation in the WT–t projection at the level 0.5 WT_{max}.

For CWT of AE signals that accompany the ductile or ductile–brittle fracture (Fig. 3.9b), WT_{max} are much smaller than for brittle one. The corresponding frequency band is significantly wider, and the duration of radiation is shorter. The frequency f_{max}, at which WT_{max} is reached under brittle fracture (Fig. 3.9b, $f_{max} = 312$ kHz), is less than under ductile or ductile–brittle fracture (Fig. 3.9a, $f_{max} = 360$ kHz).

The obtained results agree well with the theoretical regularities established in Chap. 2: for the same defect size, the shear crack opening displacement (ductile fracture) generates elastic waves of higher frequency than for the normal crack opening displacement (brittle fracture). With increasing crack size, the frequency bandwidth narrows, and the value of the dominant frequency decreases regardless of the mechanism of defect formation [44].

It is known that under the fracture of structural materials, a number of AE events can occur in a short period of time, which overlap on the combined stress–strain and AE activity curve. Therefore, a small part of such a diagram for hardened carbon steel (Fig. 3.10) is selected, and individual AE events (in Fig. 3.10 denoted by numbers *1–5*) are analyzed, calculating the energy parameter E_{WT} of the local maxima of the CWT of AE signals (Table 3.3).

It is found that during material fracture according to the energy criterion, the different types (ductile–brittle, brittle, ductile) alternate, which confirms the basic principles of fracture mechanics on the formation of plastic zones, fracture of inclusions, jump-like crack propagation. Obviously, at each stage of crack propagation, these mechanisms are present.

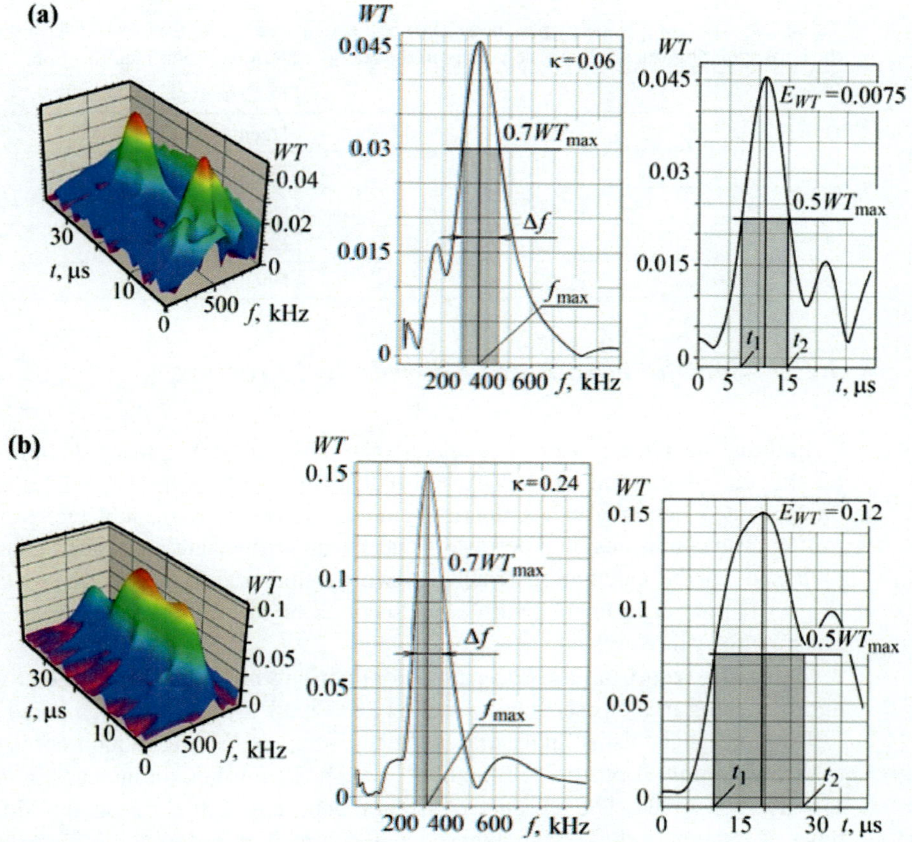

Fig. 3.9 CWT of AE signals for ductile **a** and brittle **b** fracture (3D plot; the 2D-frequency and 2D-time projections at the moment of reaching the value WT_{max}, respectively)

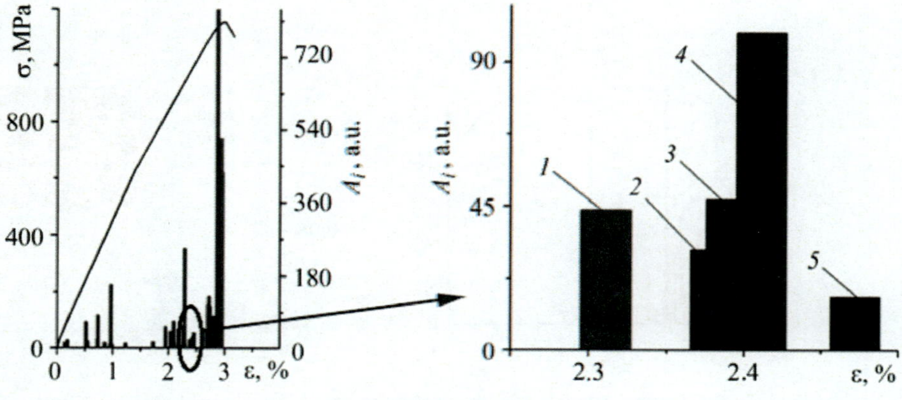

Fig. 3.10 Separate part of the stress–strain curve and distribution of AE signal amplitudes by samples under fracture of hardened carbon steel samples

Table 3.3 Value of the energy parameter E_{WT} of the local maxima of the CWT of AE signals and the corresponding fracture types according to the energy criterion of their identification

No	E_{WT}	Fracture type
1	0.045	Ductile–brittle
2	0.015	Ductile–brittle
3	0.065	Ductile–brittle
4	0.12	Brittle
5	0.009	Ductile

3.2.2 *Specific Features of Alloyed Steel Fracture*

Cylindrical specimens with a diameter of 4 mm and a length of 44 mm made of alloyed steel (0.38C-0.18 V-0.35Si-0.3Cu-0.35Mo-0.25Mn-3Ni-0.025P-1.2Cr-0.025S wt%), hardened at a tempering temperature of 893 K are subjected to tensile testing. The conditions of experiments and set-up settings are mentioned above.

Based on the calculation results, a two-parameter distribution of AE signals (Fig. 3.11) is constructed for different sections of the stress–strain curve, distinguishing their three types.

As can be seen from the two-parameter distribution (Fig. 3.11b), the activity of microcracking in the early stages (section I of Fig. 3.11a) is small compared to the dislocation mechanisms and increases at the stage of loading, which corresponds to the beginning of plastic deformation (section II). Further, an increase in the AE activity is observed. The presence of AE at this time, which correspond to different types of fracture, indicates the intensive development of micro- and macrocracking. Further deformation is accompanied by local plastic shear. The last stage of loading (section III) proceeds with reduction of the deformed volume and localization of the deformation zone (the neck is formed). This process is accompanied by the growth

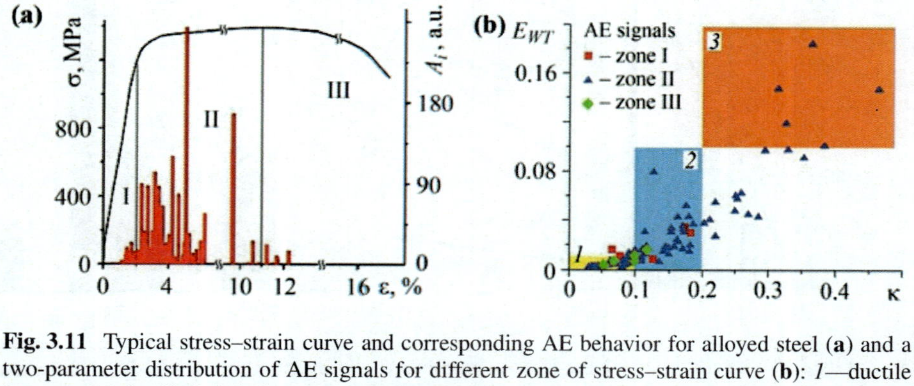

Fig. 3.11 Typical stress–strain curve and corresponding AE behavior for alloyed steel (**a**) and a two-parameter distribution of AE signals for different zone of stress–strain curve (**b**): *1*—ductile fracture, *2*—ductile–brittle fracture, *3*—brittle fracture

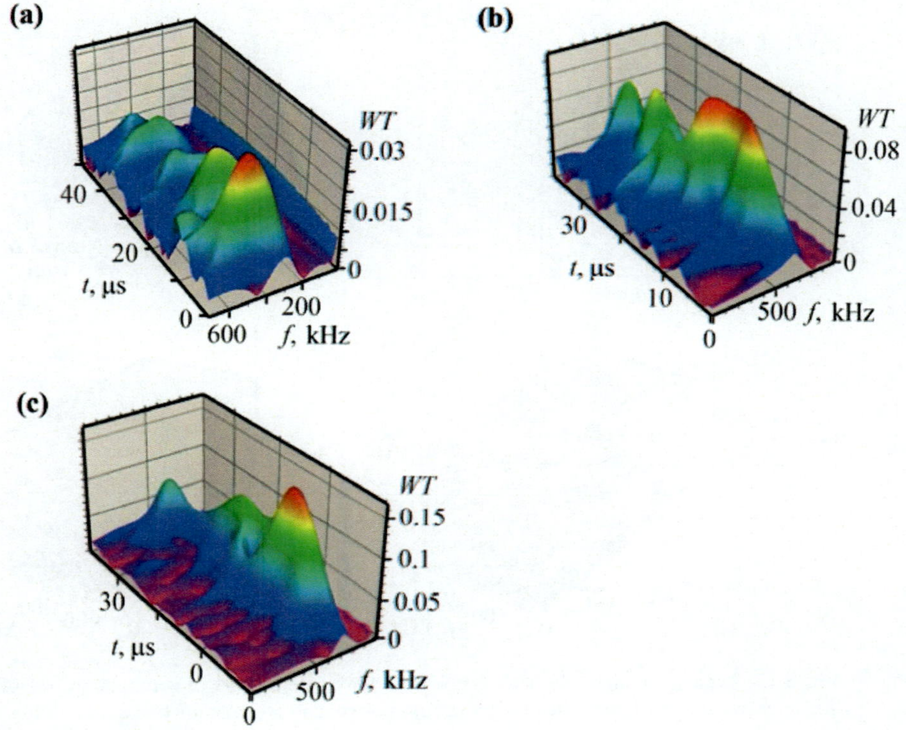

Fig. 3.12 CWT of typical AE signals for ductile **a**, ductile–brittle **b** and brittle **c** fracture of alloyed steel

and coalescence of previously formed microcracks, which ultimately leads to the sample fracture.

The values of the energy criterion parameter E_{WT} for different types of alloyed steel fracture are as follows: $0.001 \leq E_{WT} \leq 0.009$ for ductile, $0.01 \leq E_{WT} \leq 0.098$ for ductile–brittle and $0.147 \leq E_{WT} \leq 0.185$ for brittle. Figure 3.12 shows the CWT of AE signals that accompany various processes during steel fracture.

Calculations of the share of different types of fracture by energy parameter show that during the fracture of alloyed steel their largest percentage (68.3%) corresponds to ductile and ductile–brittle fracture, i.e., to the processes of dislocation and micro-cracking (sections 1, 2 in Fig. 3.11b). At the same time, for brittle hardened carbon steel, the ductile–brittle and brittle fracture dominate—98.2% (see also Fig. 3.8b).

3.2.3 *Specific Features of Brittle Materials Fracture*

Brittle materials are subjected to different loading modes: three-point bending for cylindrical samples of corundum (Ø 13.5 mm, $l = 80$ mm); compression for glass paving slab (the slab has the shape of a truncated pyramid, the lower base of which

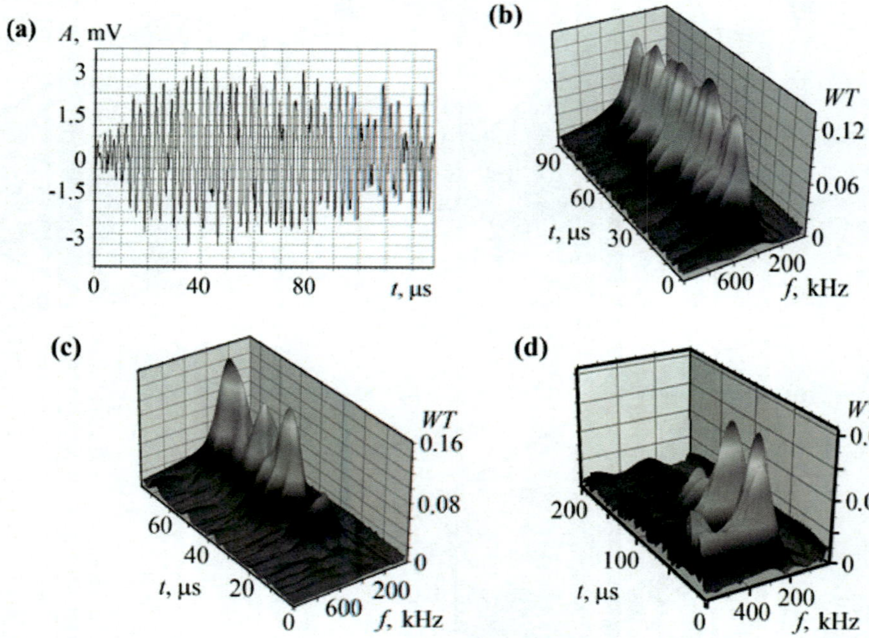

Fig. 3.13 Typical AE signal (**a**) and its CWT (**b**) for brittle fracture of corundum, CWT of typical AE signals under brittle fracture of glass paving slab (**c**) and propagation of microcracks in window glass (**d**)

is a square with a side of 95 mm, and the upper one with a side of 90 mm. The slab is hollow inside (no bottom face) with the side walls 15 mm thick and the upper face 10 mm thick; slab height is 60 mm. At last, we considered the tension of lamellar and eccentric tension of compact samples of window glass. Figure 3.13 presents the CWT of typical AE signals during brittle materials fracture.

According to the two-parameter distribution of AE signals (Fig. 3.14), the values of the energy parameter E_{WT} are in the range of 0.01–0.305, and the criterion index κ is in the range of 0.06–0.56. For most AE signals, the energy index $E_{WT} > 0.01$, and the criterion $\kappa > 0.1$ (sections 2 and 3 in Fig. 3.14).

These signals correspond to ductile–brittle and brittle fracture, i.e., they accompany the development of microcracking and propagation of macrocracks, respectively. It is found that for all studied brittle materials, the ductile–brittle fracture (corundum—45.8%, glass paving slab—19.15%, window glass—22.15%) and brittle fracture (corundum—54.2%, glass paving slab—80.85%, window glass—77.75%) prevail.

The fracture types of the investigated materials correspond to the values of the energy parameter E_{WT}. In particular, the corundum, if $0.01 \leq E_{WT} \leq 0.086$, fractures by the ductile–brittle mode and if $0.1 \leq E_{WT} \leq 0.3$ by brittle one; window glass, if $0.015 \leq E_{WT} \leq 0.076$—the fracture mode is ductile–brittle and if

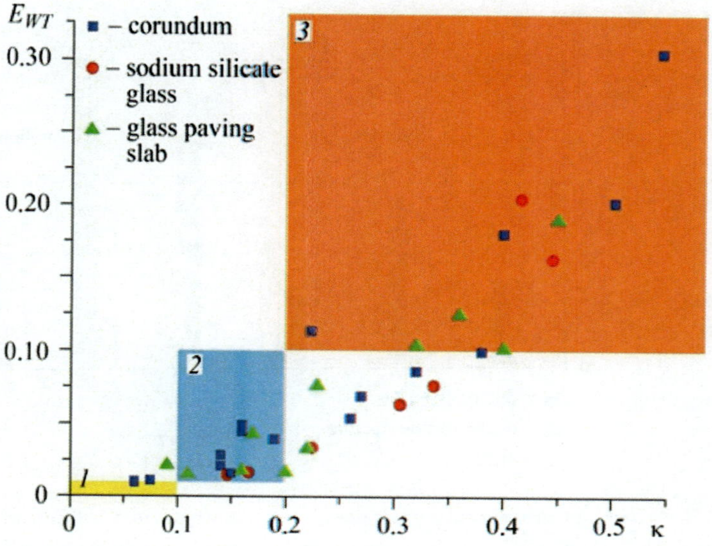

Fig. 3.14 Two-parameter distribution of AE signals during the fracture of brittle materials

$0.163 \leq E_{WT} \leq 0.215$—brittle; glass paving slab, if $0.012 \leq E_{WT} \leq 0.08$ fractures ductile–brittle and $0.102 \leq E_{WT} \leq 0.206$—brittle.

Table 3.4 presents the ranges of the criteria parameter E_{WT} values for all investigated structural materials by the types of fracture.

Summarizing the obtained results, a new energy criterion for identifying the types of macrofracture of structural materials by the energy parameter of AE signals local pulses is established:

Table 3.4 Ranges of the values of criteria parameter E_{WT} for the investigated structural materials

Structural material	Type of fracture		
	Ductile	Ductile–brittle	Brittle
Carbon steel	$0.0034 \leq E_{WT} \leq 0.0097$	$0.011 \leq E_{WT} \leq 0.047$	–
Hardened carbon steel	$0.003 \leq E_{WT} \leq 0.0098$	$0.01 \leq E_{WT} \leq 0.08$	$0.1 \leq E_{WT} \leq 0.12$
Alloyed steel	$0.001 \leq E_{WT} \leq 0.009$	$0.01 \leq E_{WT} \leq 0.098$	$0.147 \leq E_{WT} \leq 0.185$
Corundum	–	$0.01 \leq E_{WT} \leq 0.086$	$0.1 \leq E_{WT} \leq 0.3$
Window glass	–	$0.015 \leq E_{WT} \leq 0.076$	$0.163 \leq E_{WT} \leq 0.215$
Glass paving slab	–	$0.012 \leq E_{WT} \leq 0.08$	$0.102 \leq E_{WT} \leq 0.206$
By statistical analysis[a]	$E_{WT} < 0.0095 \pm 0.0004$	$0.0112 \pm 0.002 \leq E_{WT}$ $E_{WT} < 0.053 \pm 0.038$	$E_{WT} \geq 0.122 \pm 0.03$

[a] Statistical analysis of the data was performed using the ANOVA package, determining the mean values and standard deviations of the energy parameter

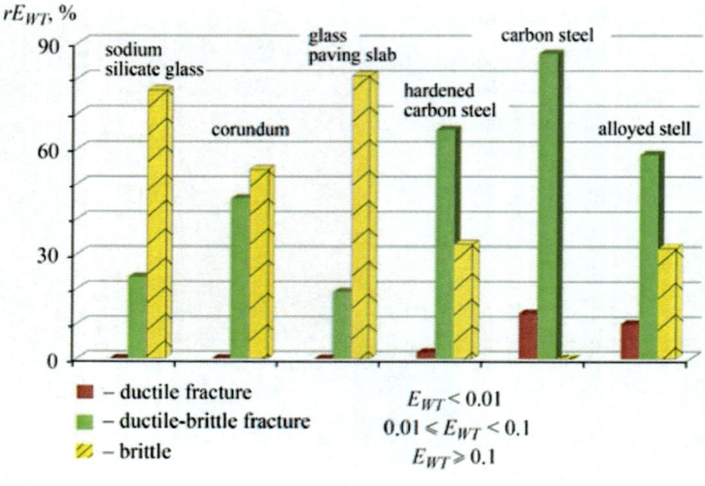

Fig. 3.15 The ratio of energy of the representative fracture types for the different structural materials

$E_{WT} < 0.01$ for ductile fracture (plastic deformation);
$0.01 \leq E_{WT} < 0.1$ for ductile–brittle (microcracking);
$E_{WT} \leq 0.1$ for brittle (propagation of the macrocrack).

Figure 3.15 shows a diagram of the distribution of different types of fracture for the studied structural materials according to the energy parameter. It is found that in the brittle materials (glass, corundum, hardened carbon steel) ductile–brittle and brittle fracture prevails, and in the ductile ones (carbon steel, alloyed steel) ductile and ductile–brittle fracture dominates. This agrees well with the previously obtained results [6, 10].

In the constructed criterion, the values of the new quantitative criteria index E_{WT} have wider limits of transition from the ductile to brittle fracture than those by the parameter κ. According to the literature [45], the extension of the numerical limits of E_{WT} for the ductile–brittle fracture, as compared to the parameter κ, increases the reliability of its identification by the energy criterion and hence the reliability of establishing other types of fracture. So the results of TD is improved in this case.

3.3 Continuous Wavelet Transform of the AE Signals Emitted Under Fracture of Aluminum and Its Alloy

3.3.1 Investigation of the Aluminum Fracture by Parameters of the CWT of AE Signals

Several physical phenomena, in particular, the motion of dislocations, shear, lattice twinning, phase transitions, which in some cases can occur in a stressed structure

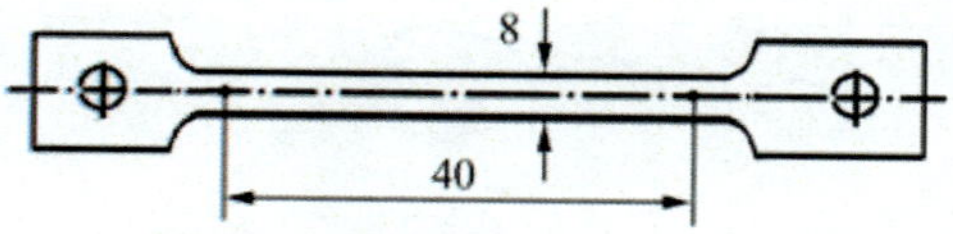

Fig. 3.16 Geometry of aluminum samples (mm)

simultaneously can be, in addition to crack initiation, the source of AE for metals. Methodologically, the AE signals from dislocations and cracks can be separated by studying the parameters of AE from materials in which one main radiation source is known in advance. For this purpose, pure aluminum (99.36%), which is well studied both in the form of monocrystals and polycrystals and alloys [46, 47], is chosen. For example, it is known [16] that in annealed pure aluminum under uniaxial compression up to 3–5% microcracks are practically not formed, and its deformation is due to the motion of dislocations.

Smooth cylindrical aluminum samples (Fig. 3.16) (99.36Al-0.319Si-0.095Ti-0.007Mn-0.241Fe-0.014Zn wt.%) are tensile tested in the as-supply state and after annealing with the help of the UME-10TM (Russia) loading unit at $t = 415$ °C.

Five samples of each type of material are tested repeatedly. Simultaneously with the tensile testing, the AE information is recorded by the SKOP-8M system. The modes of AE measurements are like those mentioned above (see subsection 3.2).

According to the strain–stress curves (Fig. 3.17) aluminum belongs to ductile materials. After annealing, the activity and intensity of AE radiation under aluminum tensile testing decreased slightly (Fig. 3.17b) compared to the as-supply state (Fig. 3.17a), as evidenced by the decrease in the number of AE events and their total amplitude. It is known that under plastic deformation of aluminum, the activity of AE coincides with the density of mobile dislocations [48, 49]. Thus, it can be concluded that after annealing of the aluminum, the number of mobile dislocations decreases due to the removal of non-equilibrium structures and deformation defects. The metal becomes more plastic, and during its fracture, the ductile mechanisms, that generate AE with lower amplitudes, to compare with a brittle fracture, prevail.

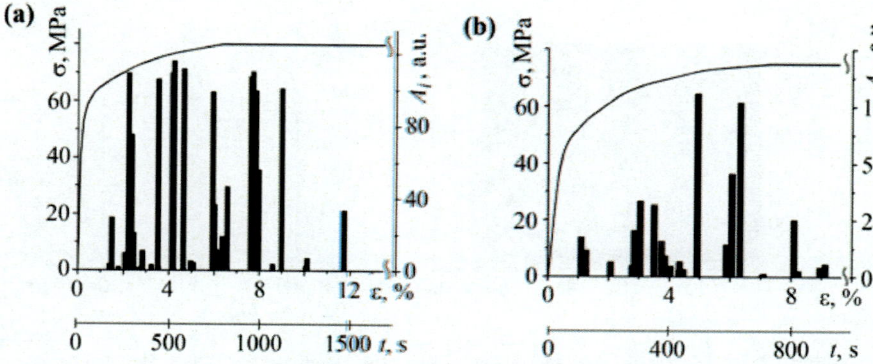

Fig. 3.17 Typical stress–strain curves and distribution of the AE signal amplitudes by samples during fracture of pure aluminum samples: **a** in the as-supply state; **b** after annealing

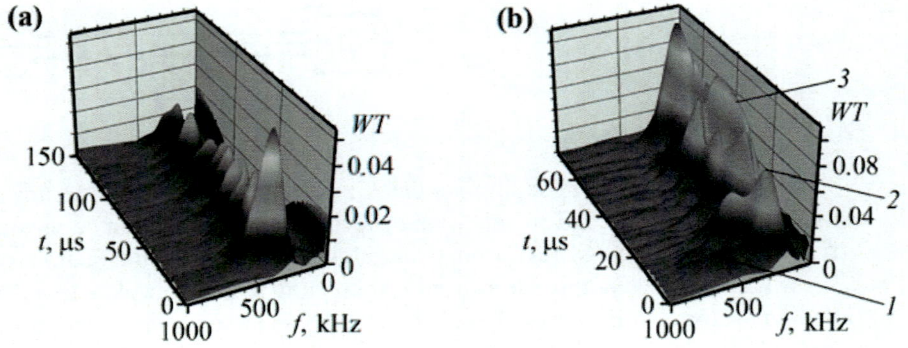

Fig. 3.18 CWT of typical AE signals under tensile testing of aluminum in the as-supply state: **a** with one local peak; **b** with several peaks at different frequency ranges

Figure 3.18 shows the CWT of typical AE signals under tensile testing of the aluminum samples in the as-supply state. Similar signals are observed during the samples testing under tension after annealing.

The CWT parameters of the signal presented in Fig. 3.18a, are as follows: $WT_{max} = 0.05$, $f_{max} = 330$ kHz, $\Delta f = 190$ kHz, $\Delta t = 10$ μs, $E_{WT} = 0.019$. This allows us to identify the ductile–brittle fracture or microcracking by the energy criterion. Figure 3.19a shows the projection of the CWT of AE signal from Fig. 3.18b on the WT–*f* plane and the scheme for determining the parameters of local pulses WT_{max}, f_{max}, Δf and Δt, the values of which are given in Table. 3.5 for each local event in the AE signal (denoted by numbers *1*–*3*).

The energy parameter E_{WT} of the local pulse is calculated by Eq. (3.3), and the type of fracture is determined according to the constructed criterion.

It is found that prior to the microcrack growth (local event *3*), plastic deformation (events *1* and 2) occurs in the prefracture zone, which is registered, apparently, due to significant concentration of the dislocation clusters under tensile loading [50].

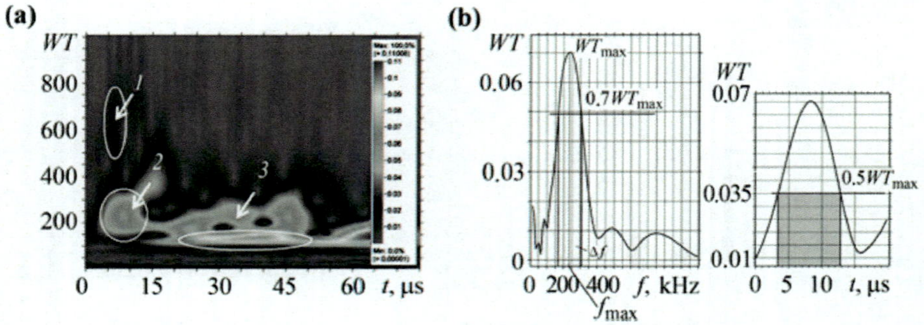

Fig. 3.19 Projection of CWT of AE signal from Fig. 3.18b on the WT–*f* plane (**a**); the scheme of determining the parameters of local pulses WT_{max}, f_{max}, Δf and Δt (**b**)

Table 3.5 Values of CWT parameters of the AE signal local peaks (Fig. 3.18b) and the corresponding fracture types under tensile testing of aluminum sample in the as-supply state

AE event	WT_{max}	f_{max}, kHz	Δf, kHz	Δt, μs	E_{WT}	Type of fracture
1	0.0015	600	300	4	0.0021	Ductile
2	0.07	230	160	10	0.0044	Ductile
3	0.08	140	50	20	0.105	Ductile–brittle

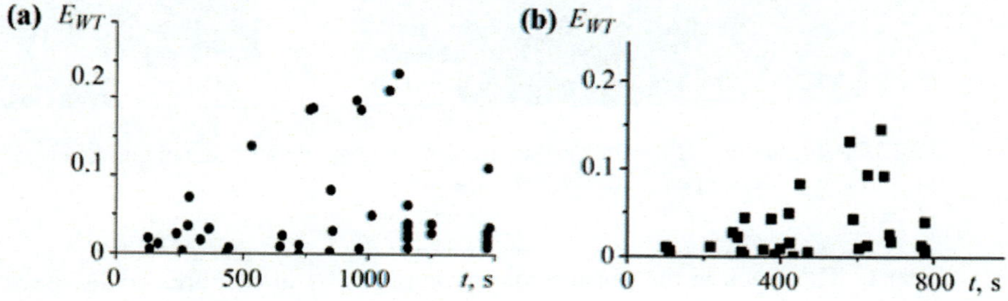

Fig. 3.20 Time distribution of AE signals by the energy parameter E_{WT} during aluminum samples tensile testing in the as-supply state (**a**) and after annealing (**b**)

Figure 3.20 illustrates the time distribution by the energy parameter E_{WT} of AE signals during aluminum samples tensile testing in the as-supply state (Fig. 3.20a) and after annealing (Fig. 3.20b).

As one can see, during the time interval of loading in both cases, AE signals are mainly generated, which according to the energy criterion accompany the ductile and ductile–brittle fracture. After 500 s from the beginning of the experiment for the aluminum in the as-supply state and 600 s after annealing, AE signals appear. This characterizes brittle fracture, i.e., the coalescence of microcracks and the transition to the macrocrack propagation. Note also that during the sample tensile testing in the as-supply state, the values of the energy parameter of these signals are slightly higher than under tension after annealing, which confirms the higher plasticity of the latter.

The obtained results on identifying the types of fracture agree well with the known fractographic studies of the fracture surfaces of aluminum samples [51]. On the fracture surface of smooth cylindrical specimens ruptured under tension, three characteristic zones are most often distinguished: fibrous, radial, and shear (Fig. 3.21). The main area of the fracture surface is occupied by the fibrous zone formed by the fracture pits (Fig. 3.21b).

The arrows indicate fibrous thin lines near the pits and on their inner surfaces, which testify to the pits increase due to material sliding along different systems. This area is also characterized by randomly located fibers. In general, the fibrous area which corresponds to the slow crack growth, is concentrated in the center and surrounds the center of fracture, which is located mainly on the axis of tension or

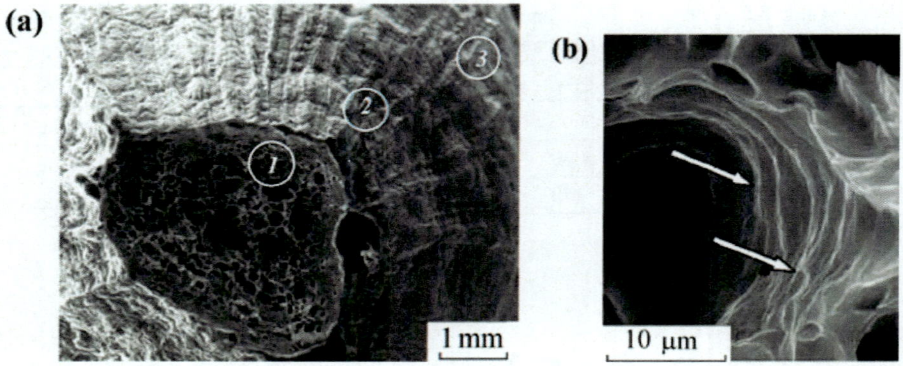

Fig. 3.21 Fractogram of the fracture surface of commercially pure aluminum: **a** general view of the surface; **b** fracture pits

near it. The crack in the fibrous zone propagates by micropores coalescence due to the fracture of the walls between them.

During the crack transition from the stage of slow growth to the stage of rapid or unstable propagation, radial scars are formed. The direction of the scars trend coincides with the general of the crack propagation. The cut area consists of an even circular region adjacent to the free surface of the sample. Many sliding bands are recorded on the lateral fracture surface of the samples near the fracture surface, which confirms the conclusion about the material plasticity.

The results of the investigation of kinetics of the microcracks initiation and propagation under plastic deformation of aluminum are presented in [52]. Here, the microscopic cracks are concentrated on the slip traces. Each crack is oriented perpendicular to the direction of the applied maximum tensile stresses, regardless of the orientation of the slide trace. Under significant deformations, between these microscopic cracks, the cracks sometimes appear, directed along the slide trace (Fig. 3.22). The results of the research of a few authors who study slide lines in aluminum also indicate that the visible shear lines in some cases consist of separate extremely thin microcracks.

Fig. 3.22 Scheme of the microcrack initiation and propagation in aluminum during plastic deformation

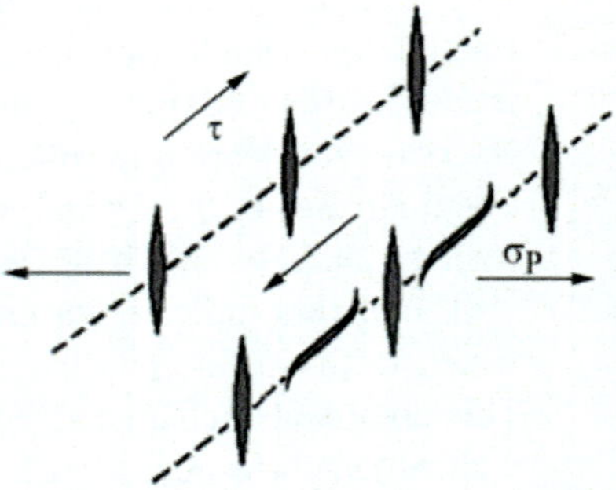

Thus, the fracture of commercially pure aluminum is ductile with the formation on the fracture surface of many slipbands and deep pits, accompanied by the generation of AE signals with the corresponding energy parameters.

3.3.2 Analysis of AE Signals During Fatigue Fracture of Aluminum Alloy

For effective AE-diagnostics of the state of structural elements made of aluminum alloys, it is important to study the AE activity and peculiarities of signals during the initiation and development of fatigue failure in them.

The initiation of fatigue fracture is investigated by cantilever bending testing of prismatic beam specimens of dimensions 10 × 20 × 160 mm made of heat-hardened aluminum alloy, grade 1201-T (93.2Al-0.25Mn-6.2Cu-0.15Zr-0.08Ti-0.1 V) [7].

The alloy represents the Al–Cu–Mn alloying family, heat-hardened and artificially aged. According to the literature [53], the alloy microstructure consists of grains of α-solid solution of copper and manganese in aluminum, which are strengthened by the inclusion of Al_2Cu and $Al_{12}Mn_2Cu$ secondary phases, which are uniformly distributed in the grain and along its boundaries in the form of flakes. The fatigue crack tip propagates mostly along the local clusters of brittle intermetallic phases and grain boundaries. The fatigue fracture surface of the alloy (Fig. 3.23) consists of brittle fatigue grooves with a significant number of quasi-cleavage facets, thus indicating a brittle fracture mechanism.

Figure 3.24 shows a typical AE signal (Fig. 3.24a) and its CWT (Fig. 3.24b) under AE testing of the alloy.

The CWT parameters of the AE signals are ranged as: $WT_{max} = 0.006–0.15$; $f_{max} = 250–330$ кHz, $\Delta f = 100–120$ kHz, $\Delta t = 10–15$ μs, and the criteria parameter $\kappa = 0.25–0.5$, calculated by formula (1.9), what corresponds to the brittle fracture. Such signals are apparently generated due to the fracture of the thermally hardened solid solution of copper and manganese in aluminum, which is the basis of the alloy microstructure, and the quasi-brittle cracking of coarse inclusions of Al_2Cu

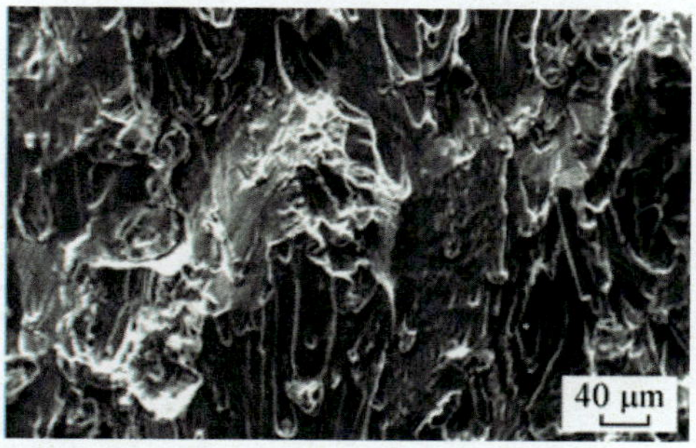

Fig. 3.23 Microfractogram of fatigue fracture surface of 1201-T alloy

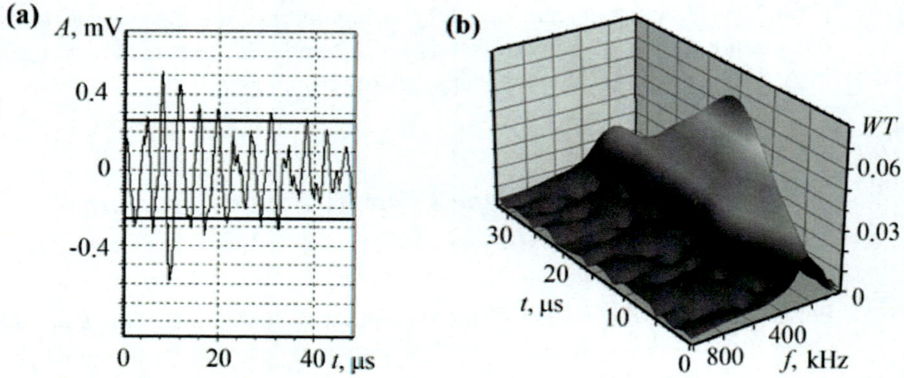

Fig. 3.24 Waveform (**a**) and CWT (**b**) of a typical AE signal during fatigue fracture of 1201-T alloy

intermetallic components concentrated at the grain boundaries. Energy parameter E_{WT} of AE signals is in the range of 0.15–0.32.

As it is known [46], there are two sources of AE in aluminum alloys: (1) plastic deformation (propagation of dislocations, tear of dislocation segments from fixing points, etc.); (2) fracture of intermetallic components distributed in the matrix. Unfortunately, it is not possible to separate the AE signals caused by plastic deformation of the alloy by the κ-criterion.

It is established, by calculating the energy parameter E_{WT} of local CWT maxima, that during the alloy fatigue fracture, the AE signals are generated, which accompany the matrix plastic deformation. For them $E_{WT} = 0.004–0.006$ (Fig. 3.25).

Thus, it is found by the energy criterion that during fatigue loading, a brittle fracture of the 1201-T alloy occurs, which alternates with ductile one and the development of microcracking, what is proved by fractographic studies.

Fig. 3.25 Distribution of different types of 1201-T alloy fracture by κ and energy criteria

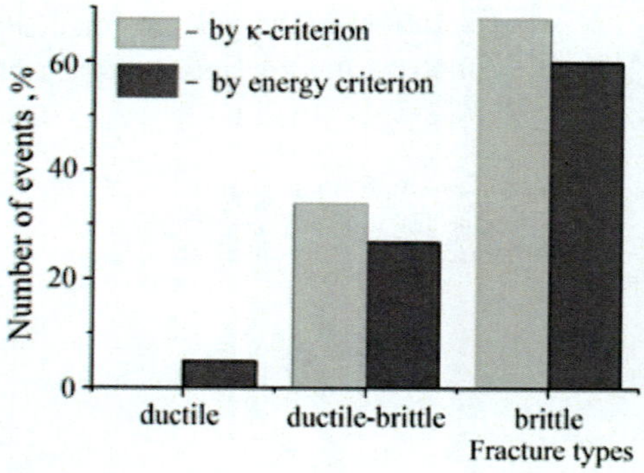

3.4 Specific Features of the Acoustic Emission Signals During Fracture of Aluminum Alloy Welded Joints Under Quasi-Static Loading

The main elements of metal structures are connected with the help of welded joints. Their quality determines the operational reliability and economy of the structure as a whole [54]. Residual deformations and stresses due to uneven heating and cooling of the metal during welding and significant stress concentrations facilitate the development of various defects in them. This leads to seal failure, reduction in strength, and deterioration of other performance properties of structural elements under repeated static, alternating cyclic, dynamic and vibration, etc. loads. Therefore, during TD, considerable attention is paid to the study of the state of the welded joints, as the areas of potential development of fracture.

Aluminum and its alloys are the most common structural materials in many industries. There are published studies of the welded joints of various aluminum alloys by the AE method [8, 55–60]. To study their fracture character, such AE parameters are used: total count [56, 58, 59], AE activity [57], frequency parameters of CWT, and the sum of the amplitudes of AE signals [8, 60].

Because of the structural and mechanical inhomogeneity, each of the welded joint zones (base metal, heat affected zone (HAZ), fusion boundary, weld metal) of the aluminum alloy causes during fracture the generation of AE signals with some characteristic features.

Prismatic samples of thermally hardened 1201-T aluminum alloy with dimensions of $10 \times 20 \times 160$ mm are subjected to the three-point bending testing. The samples are prepared of the welded joints of plates (thickness 20 mm), using the trough electron beam welding without a filler metal, with an induced fatigue crack: crack of Type I—in the base metal; II—in the HAZ, III—in the fusion zone and IV—in the weld metal [8]. The conditions of experiments and set-up settings are mentioned above (see subsection 3.2).

3.4.1 *Identification of Fracture Sources of Different Zones of Aluminum Alloy Welded Joints by Fractographic Analysis of the Fracture Surfaces*

As noted above and in [53], the microstructure of the alloy at room temperature consists of the grains of α-solid solution of copper and manganese in aluminum and the Al_2Cu secondary phase.

The fracture surface of the base metal (Fig. 3.26) is of the brittle–ductile type. The fracture surface consists of large bulges and dents, the surface of which is covered with small quasi-cleavage facets. Comparing the microstructure and fractogram of the alloy base metal, it is found that the shape and size of the grains coincide with the dimensions of the buldges and dents and are approximately 120–150 μm. The

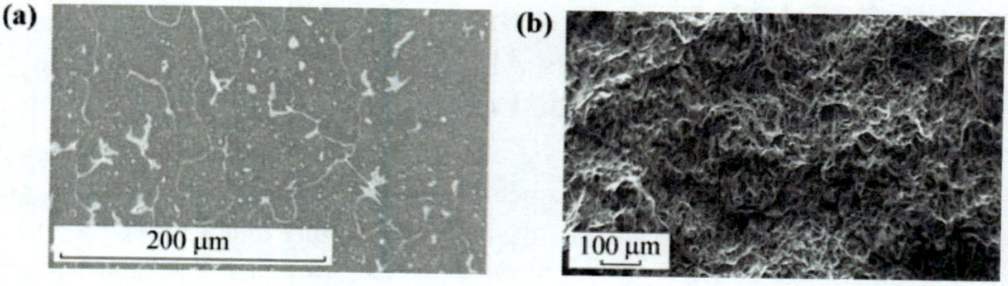

Fig. 3.26 Microstructure (**a**) and fractogram of the fracture surface (**b**) of 1201-T alloy

geometry of the inclusions of the Al_2Cu secondary phase on the microstructure is commensurable with the size of the fragments on the microfractogram (25–35 μm).

Thus, in the base metal of the alloy under static loading, the crack propagates in most cases outside the grains, the character of the fracture is brittle–ductile, in which the ductile component corresponds to the fracture of α-solid solution, and the brittle one—to cracking of inclusions of the strengthening phase.

The microstructure of the HAZ consists mainly of the recrystallized and copper-depleted grains. Copper, as a secondary phase of Al_2Cu, is released at their boundaries and coagulates in the form of local aggregations during the redecomposition of the solid solution (Fig. 3.27a).

The metal in this area is somewhat more ductile than the base one, so the ductile type of fracture predominates on the fractogram, although in some places facets of the brittle inclusions cracking are visible (Fig. 3.27b).

At the fusion boundary between the HAZ and the weld metal, the dimensions of structural elements change, the coarse melted recrystallized grains are connected with the fine-dispersed metal of the welded joint (Fig. 3.27c). Intermetallic phases accumulate at the fusion boundary, which causes a decrease in its plasticity. The fracture surface can be characterized as macrobrittle with quasi-cleavage facets of large eutectic inclusions cracking (Fig. 3.27d).

As a result of the high cooling rate, the structure of the weld metal is fine-grained (Fig. 3.27e). The fracture is ductile with a cup-shaped relief and sites of plastic flow of the metal (Fig. 3.27f).

3.4.2 *Peculiarities of AE Signals During Fracture of the 1201-T Alloy Welded Joint*

To study the dynamics of fracture of different zones of the welded joint, the energy criterion for identifying the types of fracture is used.

Analyzing the AE signals during fracture of the alloy base metal by the κ-criterion [8], two their types are distinguished: type I for $\kappa = 0.2$–0.3; type II for $\kappa = 0.5$–0.6. They correspond to the fracture of thermally hardened solid solution of copper and manganese in aluminum and quasi-brittle cracking of intermetallic compounds,

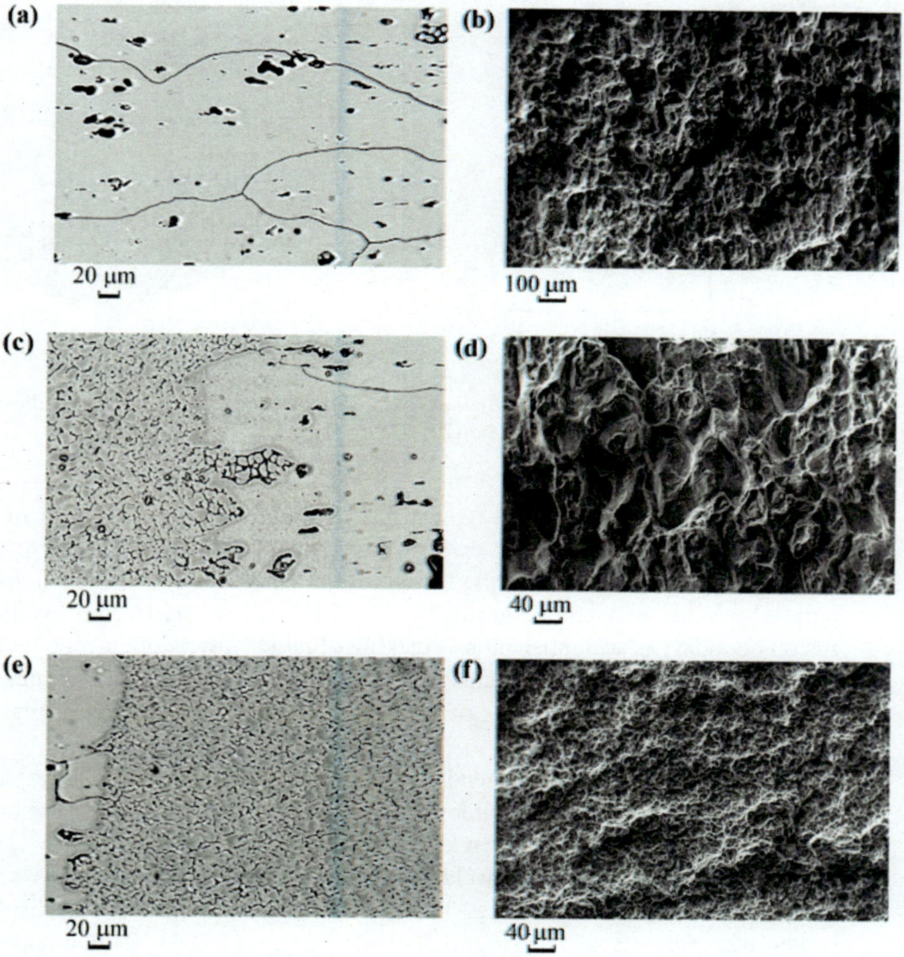

Fig. 3.27 Microstructures (**a**, **c**, **e**) and fractograms (**b**, **d**, **f**) of HAZ (**a**, **b**), fusion boundary (**c**, **d**), and weld metal (**e**, **f**) of the welded joint of the 1201-T alloy

respectively. After calculating the energy parameters of the local maxima of the CWT of AE signals by formula (3.3), three types of AE signals are established: type I—$E_{WT} = 0.0032–0.056$; type II—$E_{WT} = 0.012–0.098$; type III—$E_{WT} = 0.12–0.15$. This is in compliance with the energy criterion accompany ductile fracture (matrix fracture), ductile–brittle (quasi-brittle solid solution fracture) and brittle fracture (cracking of large inclusions of intermetallic components at grain boundaries, macrocrack propagation).

As above, using the energy criterion during the HAZ metal testing, it is possible to extract the AE signals with $E_{WT} = 0.0023–0.005$, which accompany the ductile fracture of the matrix. Most AE ($E_{WT} = 0.011–0.072$) signals are generated during the fracture of the solid solution consisting of recrystallized grains of the HAZ, depleted in the secondary phase, which coagulate at their boundary. Figure 3.28 shows the

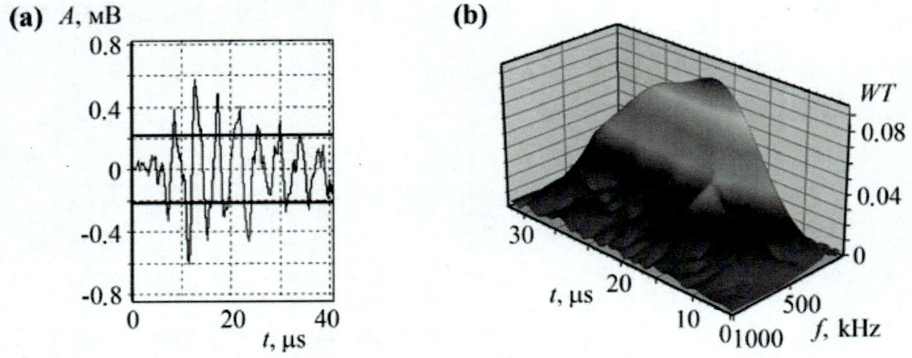

Fig. 3.28 Waveform (**a**) and CWT (**b**) of a typical AE signal resulting from intermetallic cracking during fracture of the HAZ metal of the 1201-T alloy

wave reflection and CWT of a typical AE signal, generated as a result of quasi-brittle cracking of local clusters and fracture of thin layers of Al_2Cu intermetallide at intergranular boundaries during testing of the HAZ metal. For such signals, the parameter value is in the range of E_{WT} = 0.13–0.256. The fracture of the weld fusion boundary is accompanied by cracking of large clusters of eutectic inclusions and separation of the melted grains of the HAZ. In the latter case, AE signals with an energy parameter in the range of E_{WT} = 0.142–0.390 are generated (Fig. 3.29a, b).

The brittle cleavage of intermetallic compounds at the fusion boundary generates powerful signals of high amplitudes (Fig. 3.29c, d), and the index in the range of E_{WT} = 0.325–0.564 is the highest here.

As noted above, the alloy weld has a fine-grained structure and is softened compared to the base metal (Fig. 3.27e, f). The fracture is ductile and is accompanied by weak AE signals (Fig. 3.30a) with E_{WT} = 0.0066–0.0092 and several CWT peaks in a short period of time (10 μs) (Fig. 3.30b), thus indicating almost simultaneous plastic failure of the matrix and delamination of inclusions.

So, according to the energy criterion, the cracking of the inclusion of intermetallic compounds is brittle ($E_{\mathrm{WT}} \geq 0.12$) with the maximum values of the criteria parameter E_{WT} at the fusion boundary, since here there are the largest accumulations of large intermetallic compounds. A solid solution of copper in aluminum fails mostly ductile–brittle: ductile fracture is accompanied by AE signals with the parameter E_{WT} = 0.0032–0.0056, the microcracks propagation in the matrix and the delamination of inclusions—with E_{WT} = 0.012–0.098. Outside the weld, brittle fracture prevails, as the hardness of the base metal is much higher than that of the weld and the HAZ.

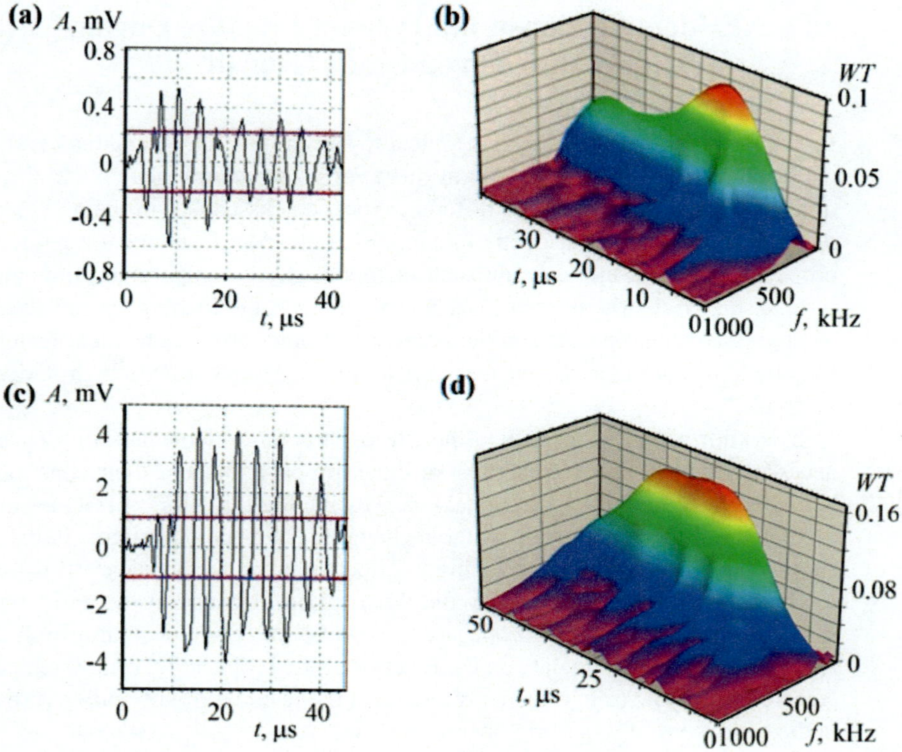

Fig. 3.29 Waveforms (**a**, **b**) and CWT (**c**, **d**) of typical AE signals during fracture of fusion boundary of the welded joint of 1201-T alloy

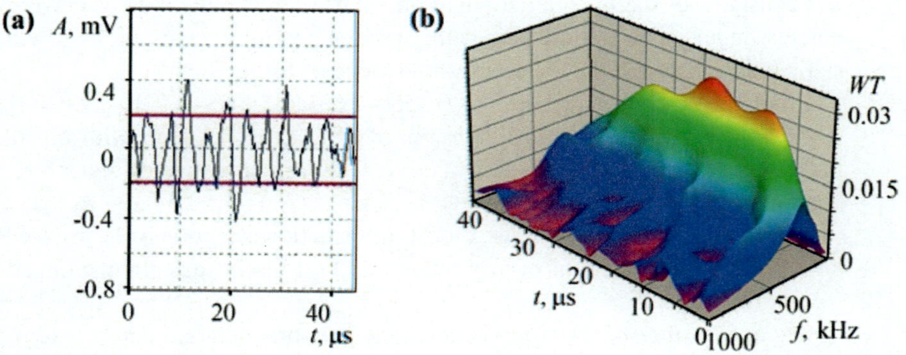

Fig. 3.30 Waveform (**a**) and CWT (**b**) of typical AE signal during fracture of weld metal of the welded joint of 1201-T alloy

3.5 AE-Identification of the Types of Fracture During Low-Temperature Creep Crack Growth

Investigation of materials creep is topical to substantiate the strength and reliability of structures [61]. During long-term operation of structural elements in the conditions of continuous loading of different types, cracks appear in them. The growth of these cracks due to the creep of materials can lead to the element fracture. Crack propagation has an influence on such factors as stress, temperature, time and type of load, material structure, etc. During long-term static loading (in the absence of an aggressive environment and the influence of other physicochemical factors), the fracture of metals occurs by the mechanisms of low-temperature or high-temperature creep [62].

It is known that to determine the life of structural elements with cracks, it is necessary to have kinetic diagrams of crack propagation. For their construction it is important to identify as accurately as possible the initiation of fracture in structural materials because it starts without changing the macrodescriptors of their state. As practice has shown, it is possible to effectively record the stages of dislocation mechanisms of microcracking by the AE method, because in the early stages of subcritical growth the crack propagates jump-like, and each local jump emits elastic AE waves. The obtained data on the kinetics of the propagation of low-temperature creep cracks allow us to effectively calculate the residual life of bodies with cracks in the conditions of long-term operation under static loading [63–65].

In [13], using experimental studies with the help of the AE method, the lower threshold value of SIF, at which the propagation of the low-temperature creep crack starts, is determined. The experiments are performed in several stages: the AE constants of the material are determined to calculate the area of newly formed defects under quasi-static loading; the low-temperature creep crack growth under static load is studied; the lower threshold value of SIF is found.

Prismatic carbon steel (0.3C–0.35Si–0.65Mn–0.04S–0.03P–0.28Cr–0.25Ni–0.45Cu wt.%) specimens with dimensions of $6.8 \times 110 \times 15.8$ mm with an artificially induced fatigue crack with a length of $l = 7.97$ mm are investigated under three-point bending testing.

During loading of the samples, the AE information is recorded also by the SKOP-8M system. The conditions of experiments and set-up settings are mentioned above (see subsection 3.2).

At the first stage of the experiments in the post-processing mode, a loading curve with the distribution of the AE signals amplitudes during fatigue crack extension is constructed (Fig. 3.31). AE signals are analyzed by the criteria for identifying the material fracture types.

Two types of AE signals are established by the κ-criterion: Type I—for ductile–brittle fracture ($\kappa = 0.1$–0.2); Type II—for the brittle one ($\kappa = 0.3$–0.4).

Creep macrocracks propagate by initiation, propagation, and coalescence of pores in the plastic zone at the crack tip [65]. The AE signals that accompany these processes are selected by the energy criterion for identifying the fracture types.

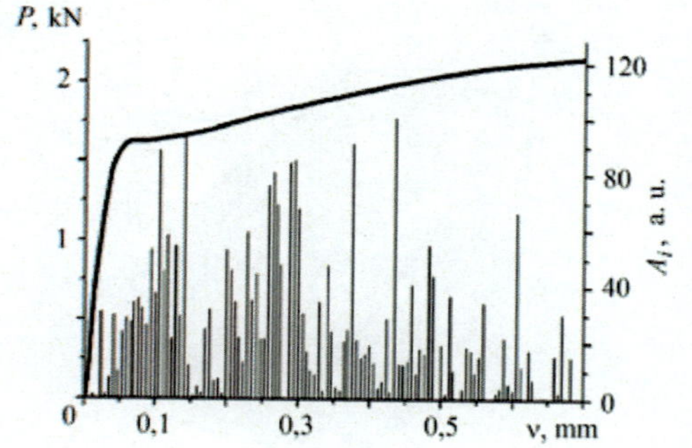

Fig. 3.31 "Load P–crack opening v" curve and distribution of AE signal amplitudes A_i

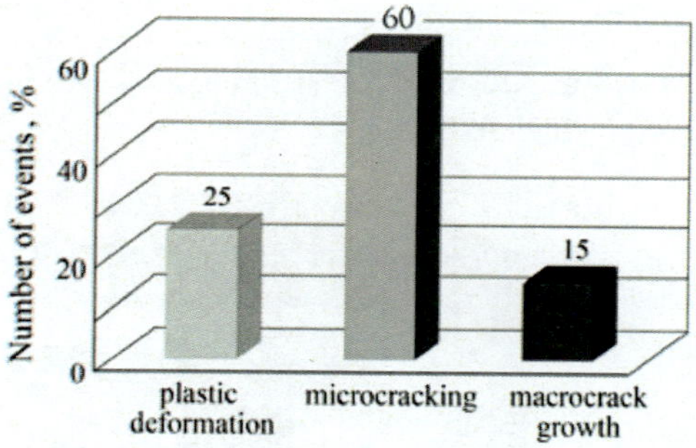

Fig. 3.32 Distribution of different fracture types under testing of carbon steel samples with fatigue crack under quasi-static three-point bending

For them, the parameter $E_{\mathrm{WT}} = 0.0034$–0.0097. For AE signals that accompany microcracking, $E_{\mathrm{WT}} = 0.019$–0.049, and for the growth of local macrocracks, $E_{\mathrm{WT}} = 0.122$–0.616. Figure 3.32 illustrates the distribution of fracture types of carbon steel during experiments.

So, according to the energy criterion, three types of fracture of carbon steel are established. The mechanisms of ductile fracture and microcracking prevail (totally 85% of the registered AE signals) among them. Figure 3.33 shows the CWT of the AE signal, which has two maxima at different frequencies (denoted by *1, 2*), thus indicating a change in the AE generation mechanisms in a short time interval (5 μs). Local event *1* at time 16.75 μs, with the CWT parameters $WT_{\max} = 0.04$, $f_{\max} = 490$ kHz, $\Delta f = 200$ kHz, $\Delta t = 6$ μs, $E_{WT} = 0.0058$, according to the energy criterion, accompanies a plastic deformation. Local event *2* at time 19.75 μs, with the parameters $WT_{\max} = 0.11$, $f_{\max} = 160$ kHz, $\Delta f = 90$ kHz, $\Delta t = 17$ μs, $E_{WT} = 0.122$, accompanies a brittle fracture—the local macrocrack growth.

To register the propagation of the low-temperature creep crack in the second stage of the experiments, similar carbon steel samples with induced fatigue cracks are used. Specimens are investigated under long-term static loading [63] on a special set-up (Fig. 3.34a) according to the scheme shown in Fig. 3.34b. The AE information is recorded simultaneously by the SKOP-8M system.

The experiments are performed for different values of SIF. Figure 3.35 shows typical distributions of AE signals amplitudes A_i and their sum ΣA_i for each time of the experiment.

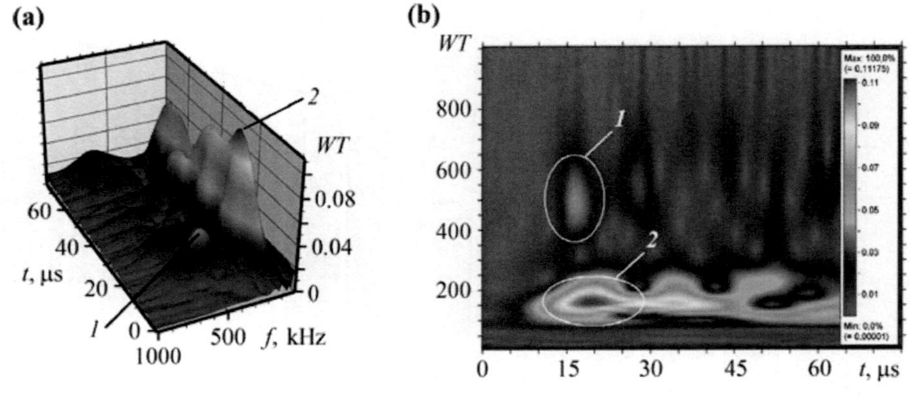

Fig. 3.33 CWT (**a**) and its projection on the WT–f plane (**b**) of AE signal during fracture of the carbon steel sample

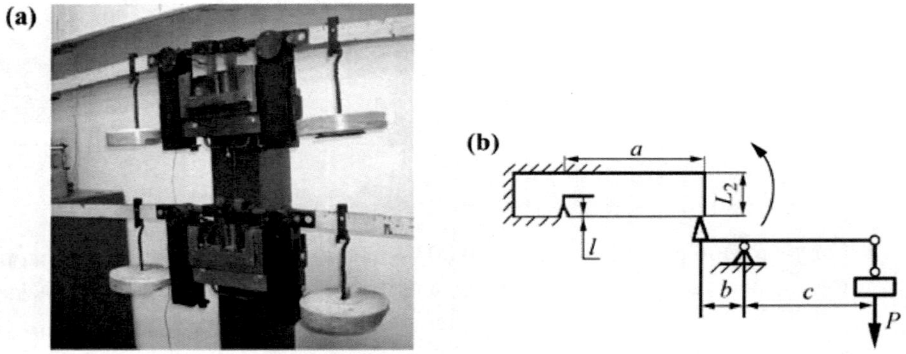

Fig. 3.34 General view of the set-up **a** and loading scheme of specimen with a crack **b** under the action of long-term static loads

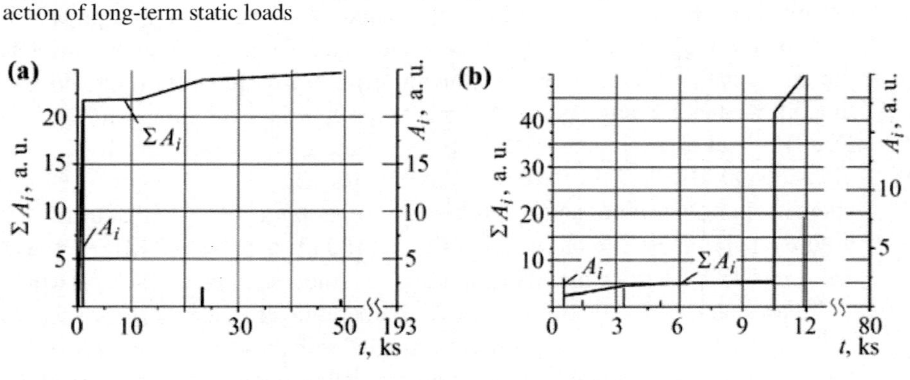

Fig. 3.35 Time distribution of AE signals amplitudes and change of their sum during an experiment for different values of SIF: **a** $K_{\mathrm{I}} = 9.98$; **b** 12.3 MPa $\sqrt{M}$

(a) (b)

Fig. 3.36 Fracture surfaces of prismatic carbon steel specimens at the initial stage of subcritical crack growth (**a**) and under the low-temperature creep (**b**)

Two types of AE signals are selected by the κ-criterion that identify ductile–brittle ($\kappa = 0.1$–0.2) and brittle fracture ($\kappa = 1$–2.4). Applying the energy fracture criterion to the AE signals, in addition to the two types mentioned above, the signals corresponding to the ductile fracture of carbon steel ($E_{WT} = 0.0015$–0.006) are selected. Energy parameter of AE signals that accompany microcracking is $E_{WT} = 0.014$–0.038, local macrocrack growth—$E_{WT} = 0.163$–0.344. The fracture surfaces of the samples confirm the results of identification of the fracture types of this steel under the action of quasi-static (Fig. 3.36a) and long-term loads (Fig. 3.36b).

Thus, the AE method can be effectively used to detect the early stages of the propagation of low-temperature creep cracks in structural elements, while identification of the types of fracture by the CWT energy parameters of AE signals allows us to correctly determine the moment of their start. This approach ensures a reliable determination of the lower threshold SIF value required to construct the first section of the low-temperature creep curve.

3.6 Application of the Wavelet Transform to Study the Features of Nonmetallic Materials Fracture

Rubber products are widely used in modern machine and instrument engineering [66]. The main features of rubber as a structural material are: a highly elastic state in a wide range of temperatures and the relaxation character of deformation. Rubber products are characterized by high abrasion resistance, gas and water resistance, chemical resistance, electric insulating properties, and low specific weight. The rubber mixture contains from 15 to 20 ingredients. These are rubber, vulcanizing agents, antioxidants, plasticizers, fillers, bulk and surface modifiers, dyes, etc.

Depending on the amount of sulfur in the rubber, different frequencies of the polymer mesh are obtained. With the introduction of 1–5% sulfur, a liquid mesh is formed, and the rubber becomes highly elastic, soft. As the sulfur content increases,

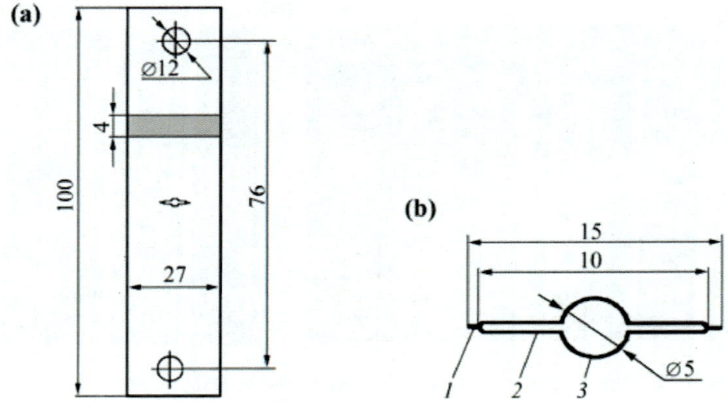

Fig. 3.37 Geometry of lamellar ebonite samples (mm) (**a**) and a crack (**b**): *1*—a crack tip with a rounding of 0.4 mm; *2*—a cut through with a thickness of 1.5 mm; *3*—a hole

the mesh structure becomes denser and the rubber—harder. When saturation of rubber with sulfur is the maximum possible (30%), a solid material is formed—ebonite. Ebonite dust, talc, pumice, white soot (3–5%) are the ebonite fillers. The resistance to the action of acids, bases, salts and fats (vegetable, animal), ensuring good electrical insulation, non-water adsorption, and gas impermeability are the important operational properties of the ebonite. Ebonite is used to produce electrical insulating parts of devices, various containers for aggressive environments, cases of acid batteries, etc. Recently, it has been practically replaced by plastics, which have better dielectric properties and chemical resistance.

Research methodology. The fracture of lamellar ebonite samples under the action of quasi-static tension is studied. Samples 4 mm thick with a central crack with a radius $\approx$ 0.4 mm are prepared according to the scheme in Fig. 3.37.

The experiments are performed on a tensile testing machine of type SVR-5 from Karpenko Physical–Mechanical Institute of the National Academy of Sciences of Ukraine with a loading speed 4×10^{-7} m/s. Test conditions and settings of measuring equipment are the same as in the previous studies.

In the post-processing mode, the fracture curves (Fig. 3.38) and CWT of the registered AE signals are constructed. It is found that already in the early stages of fracture, the measuring system registers AE signals with relatively low amplitudes, which indicates the emission of a small amount of energy.

The dominant frequency range for these local pulses is 125–250 kHz, the duration of their radiation is 10–14.5 μs, and the energy parameter is $0.016 \leq E_{WT} \leq 0.136$. They are classified as type I signals (Fig. 3.39a). According to the criterion for identifying the types of fracture, they accompany the ductile–brittle fracture, which passes to the propagation of macrocracks (brittle fracture) in the ebonite.

These signals alternate with those in which the dominant frequencies are much higher (500–600 kHz), the duration of their radiation is 5–7 μs, and the energy parameter is within $0.0016 \leq E_{WT} \leq 0.0036$. These are classified as type II signals

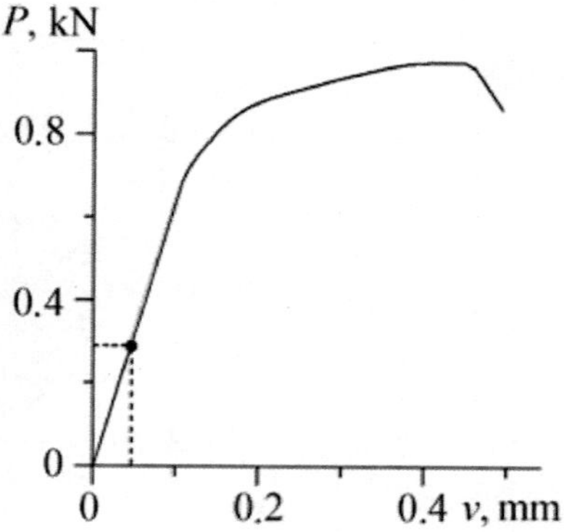

Fig. 3.38 Fracture curve of the lamellar ebonite sample in the "load *P*–crack opening *v*" coordinates (the point indicates the beginning of macrofracture)

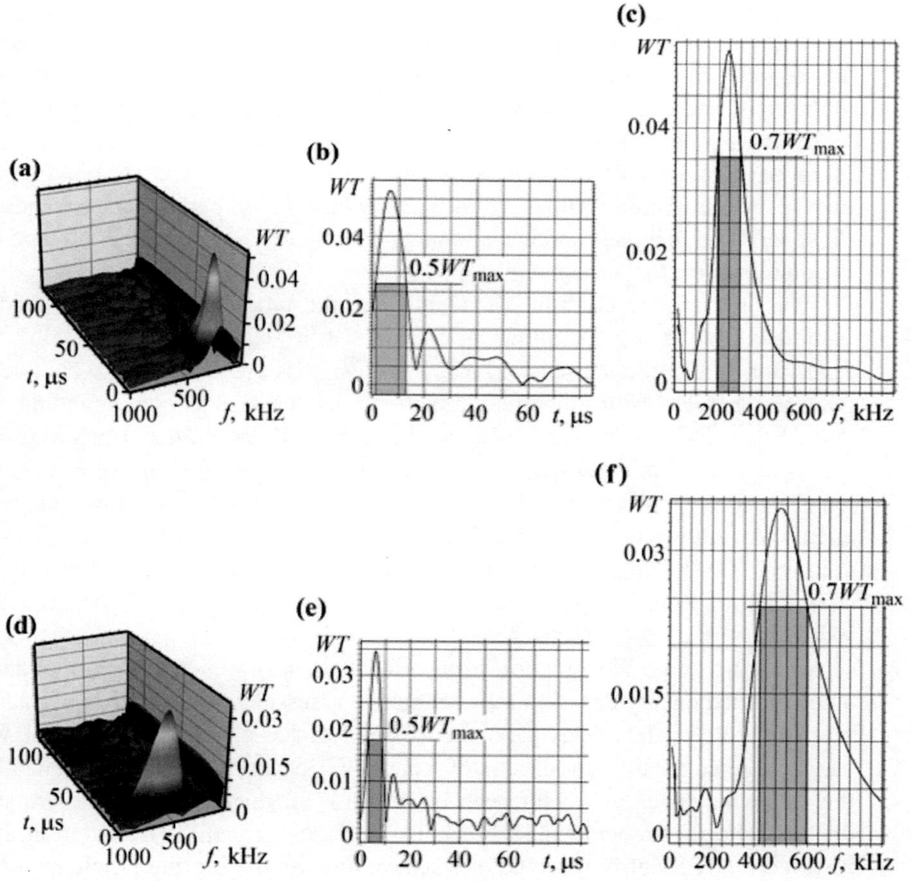

Fig. 3.39 CWT of typical AE signals under fracture of the ebonite of type I (**a**–**b**) and II (**d**–**f**): 3D plot of wavelet spectrum intensity (**a**, **d**); 2D time (**b**, **e**), and frequency (**c**, **f**) projections at the moment of reaching WT_{max}

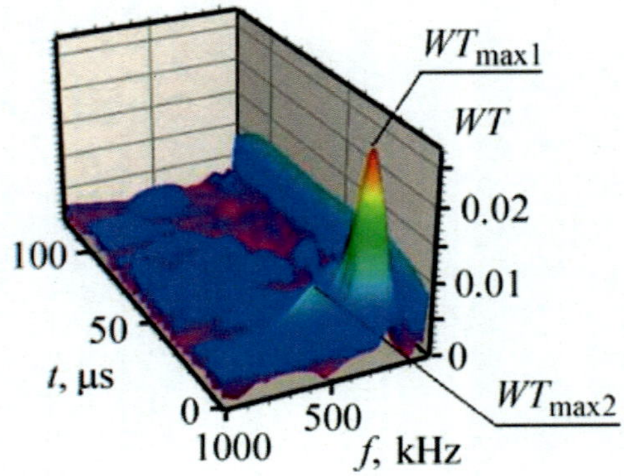

Fig. 3.40 CWT of AE signal at the final stage of the ebonite fracture

(Fig. 3.39b). Such signals according to the identification criterion accompany the ductile fracture of the ebonite.

At the end of the experiment, the AE signals with more than one spectral maximum are recorded, thus indicating a rapid and almost simultaneous development of fracture of different types in the material. In Fig. 3.40, the CWT of such typical signal is given.

In terms of the CWT parameters, the event with the value of $\mathrm{WT}_{\max 1}$ is similar to the AE signals of Type I with the energy index $E_{\mathrm{WT}} \leq 0.01$, corresponding to the ductile–brittle fracture. The event $\mathrm{WT}_{\max 2}$, whose CWT parameters coincide with those for Type II signals and whose energy parameter $E_{\mathrm{WT}} \leq 0.006$, corresponds to the ductile one.

Thus, the energy criterion for identifying the types of fracture by the CWT parameters of AE signals allows us to establish its features in nonmetallic materials. During the ebonite tension, mainly AE signals are generated, which accompany the ductile–brittle fracture, alternating with the signals corresponding to the ductile one. This can be explained by its structure. Fillers in the structure promote the formation of microcavities around nanoparticles, the fracture of which under tensile loading generates the AE signals of small amplitude and, consequently, the energy. According to our investigations, the energy parameter of such signals is in the range $0.006 \leq E_{\mathrm{WT}} \leq 0.01$. During the macrocrack initiation and propagation, as a result of which the sample is divided into two parts, the AE signals with the index $0.1 \leq E_{\mathrm{WT}} \leq 0.136$ are generated.

So, owing to the WT properties (linearity, shift, scaling, locality, normalization), a new unified energy criterion for quantitative assessment of the types of materials macrofracture by the energy parameter of CWT of AE signals is constructed, which can be applied to different classes of materials. Depending on the numerical limit, the following types of fracture can be determined: ductile (plastic deformation), ductile–brittle (microcracking), brittle (macrocrack growth). The criterion allows us to correctly identify the type of fracture, thus improving the efficiency of AE-diagnostics of structural elements.

Based on the energy criterion, the applied method for identifying the types and sources of fracture of metals and their welds has been created. During fracture of the welded joints of aluminum alloy, the solid solution of copper and manganese in aluminum fractures mostly ductile and brittle, and intermetallic compounds fail brittle with the release of the greatest energy at the fusion boundary. The most dangerous, regarding structural strength, is the transition zone from the weld metal to the heat affected zone (fusion zone), because its fracture is accompanied by AE signals with large values of the energy parameter, which characterizes its high brittleness, and, consequently, a greater probability of initiation and rapid propagation of cracks in it.

Identification of fracture types by the developed criterion with account of the energy parameters of CWT of AE signals allows us to correctly determine the start time of cracks of low-temperature creep, and, consequently, the lower threshold value of SIF required to construct the first section of the low-temperature creep curve.

References

1. Aguiar, P. R., Martins, C. H. R., Marchi, M., & Bianchi, E. C. (2012). *Digital signal processing for acoustic emission. Data acquisition applications*. Retrieved 23 March, 2021, from https://www.intechopen.com/chapters/38499
2. Anastasopoulos, A. A., & Envirocoustics, S. A. (2007). Signal processing and pattern recognition of AE signatures. In E. E. Gdoutos (Ed.), *Experimental analysis of nano and engineering materials and structures* (pp. 929–930). Springer.
3. Ono, K. (2011). Acoustic emission in materials research—A review. *Journal of Acoustic Emission, 29*, 284–308.
4. Grosse, C. U., Ohtsu, M., Aggelis, D. G., & Shiotani, T. (Eds.). (2021). *Acoustic emission testing. Basic for research—Applications in civil engineering* (2nd ed.). Springer.
5. Skalskyi, V. R., Karpash, O. M., Koshovyi, V. V., Nedoseka, A. Ya., & Stankevych, O. M. (2017). *Tekhnichna diahnostyka materialiv i konstrukcii, t 5: Akustychni metody kontrolyu dehradatsii materialiv i defektnosti elementiv konstrukcii.* [Technical diagnostics of materials and structures, vol 5: Acoustic methods of control of materials degradation and structure elements damage]. Publishing House Prostir-M, Lviv.
6. Skal'skii, V. P., Builo, S. I., & Stankevich, E. M. (2012). A criterion for evaluating the brittle fracturing of glass using acoustic emission signals. *Russian Journal of Nondestructive Testing, 48*(5), 277–284.
7. Skal's'kyi, V. R., Lyasota, I. M., & Stankevych, O. M. (2013). Acoustic-emission diagnostics of the initiation of fatigue fracture of 1201-T aluminum alloy. *Materials Science, 48*(5), 680–686.
8. Skalsky, V. R., Lyasota, I. M., & Stankevich, E. M. (2013). Peculiarities of acoustic emission signals in evaluation of fracture mechanism in welded joints on aluminium alloys. *The Paton Welding Journal, 1*, 21–27.
9. Skalskyi, V. R., Stankevych, E. M., & Basarab, R. M. (2013). Akustiko-emissionnoye diagnostirovaniye nefteperekachivayushchich stantsii (Acoustic emission diagnostics of oil pumping station equipment). *Neftegasovoye Delo (Oil and Gas Business), 11*(2), 86–90.
10. Nazarchuk, Z. T., Skal's'kyi, V. R., & Stankevych, O. M. (2014). A method for the identification of the types of macrofracture of structural materials by the parameters of the wavelet transform of acoustic-emission signals. *Materials Science, 49*(6), 841–848.
11. Skal's'kyi, V. R., Makeev, V. F., Stankevych, O. M., & Kyrmanov, O. S. (2015). Alternation of the types of fracture for dental polymers in different stages of crack propagation. *Materials Science, 50*(6), 836–843.

12. Skal's'kii, V. R., Makeev, V. F., Stankevich, O. M., Kyrmanov, O. S., Vynnyts'ka, S. I., & Opanasovich, V. K. (2015). Strength evaluation of stomatologic polymers by wavelet transform of acoustic emission signals. Strength of Materials, 47(4), 566–572
13. Skalskii, V. R., Dolinskaya, I. Y., Stankevich, E. M., & Matviiv, Y. Y. (2014). A technique for identifying the initial stage of the extension of low-temperature creep cracks. *Russian Journal of Nondestructive Testing, 50*(9), 539–547.
14. Stankevych, O., & Skalsky, V. (2016). Investigation and identification of fracture types of structural materials by means of acoustic emission analysis. *Engineering Fracture Mechanics, 164*, 24–34.
15. Kuksenko, V. S., Lyashkov, A. I., & Savelyev, V. N. (1980). Akusticheskaya emissiya pri zarozhdenii i razvitii mikrotreshchin v stalyach (Acoustic emission during initiation and development of microcracks in steels). *Defektoskopiya (Defectoscopy), 6*, 57–63.
16. Lyashkov, A. I., Inzhevatkin, I. E., & Savelyev, V. N. (1980). Izucheniye zarozhdeniya mikrotreshchin v metallach metodom akusticheskoi emissii (Study of the initiation of microcracks in metals by the acoustic emission method). *Defektoskopiya (Defectoscopy), 6*, 98–101.
17. Bunina, N. A. (1990). *Issledovaniye plasticheskoi deformacii metallov metodom akusticheskoi emissii (Study of plastic deformation of metals by the acoustic emission method).* PH of Leningrad University.
18. Takahashi, H., Khan, M. A., Kikuchi, M., & Suzuki, M. (1981). Acoustic emission crack monitoring in fracture-toughness tests for AISI-4340 and SA 533B steels. *Experimental Mechanics, 21*(3), 89–99.
19. Muravin, G. B., Finkel, V. M., Lezvinskaya, L. M., et al. (1984). Issledovaniye deformirovaniya kremnistoho zheleza metodom akusticheskoi emissii (Investigation of the deformation of siliceous iron by the acoustic emission method). *Defektoskopiya (Defectoscopy), 10*, 88–91.
20. Pardee, W., & Graham, L. J. (1975). *Analysis of acoustic emission frequency spectra.* IEEE Ultrasonic symposium proceedings, pp. 597–600. Los Angeles.
21. Mirabile, M. (1975). Acoustic emission energy and mechanisms of plastic deformation and fracture. *NDT, 8*(2), 77–85.
22. Maillet, E., Godin, N., R'Mili, M., Reynaud, P., Fantozzi, G., & Lamon, J. (2014). Real-time evaluation of energy attenuation: A novel approach to acoustic emission analysis for damage monitoring of ceramic matrix composites. *Journal of the European Ceramic Society, 34*, 1673–1679.
23. Sagar, R. V., & Prasad, B. K. R. (2011). An experimental study on acoustic emission energy as a quantitative measure of size independent specific fracture energy of concrete beams. *Construction and Build Materials, 25*(5), 2349–2357.
24. Maslov, I. I., & Gradov, O. M. (1986). Fracture energy analysis via acoustic emission. *International Journal of Fatigue, 8*(2), 67–71.
25. Muralidhara, S., Prasad, B. K. R., Eskandari, H., & Karihaloo, B. L. (2010). Fracture process zone size and true fracture energy of concrete using acoustic emission. *Construction and Build Materials, 24*, 479–486.
26. Carpinteri, A., Corrado, M., & Lacidogna, G. (2013). Heterogeneous materials in compression: Correlations between absorbed, released and acoustic emission energies. *Engineering Failure Analysis, 33*, 236–250.
27. Carpinteri, A., Lacidogna, G., & Pugno, N. (2007). Structural damage diagnosis and life-time assessment by acoustic emission monitoring. *Engineering Fracture Mechanics, 74*, 273–289.
28. Al-Balushi, K. R., Addali, A., Charnley, B., & Mba, D. (2010). Energy index technique for detection of acoustic emissions associated with incipient bearing failures. *Applied Acoustic, 71*(9), 812–821.
29. Laksimi, A., Benmedakhene, S., & Bounouas, L. (2021). *Monitoring acoustic emission during tensile loading of thermoplastic composites materials.* Retrieved 25 March, 2021, from http://www.iccm-central.org/Proceedings/ICCM12proceedings/site/papers/pap740.pdf
30. Li, Y., & Yi-Chu, Z. (2006). Wavelet analysis of acoustic emission signals from thermal barrier coatings. *Transactions of Nonferrous Metals Society of China, 16*, 270–275.

31. Khamedi, R., Fallahi, A., & Oskouei, A. R. (2010). Effect of martensite phase volume fraction on acoustic emission signals using wavelet packet analysis during tensile loading of dual phase steels. *Materials & Design, 31*, 2752–2759.
32. Chen, C., Kovacevic, R., & Jandgric, D. (2003). Wavelet transform analysis of acoustic emission in monitoring friction stir welding of 6061 aluminium. *International Journal of Machine Tools and Manufacturing, 43*, 1383-1390.
33. Piotrkowski, R., Callego, A., Castro, E., García-Hernandez, M. T., & Ruzzante, J. E. (2005). Ti and Cr nitride coating/steel adherence assessed by acoustic emission wavelet analysis. *NDT E International, 38*, 260–267.
34. Piotrkowski, R., Castro, E., & Gallego, A. (2009). Wavelet power, entropy and bispectrum applied to AE signals for damage identification and evaluation of corroded galvanized steel. *Mechanical System and Signal Processing, 23*, 432–445.
35. Physical Acoustics Corporation. (2007). *PCI-2 based AE system User's manual*. Physical Acoustic Corporation.
36. Proakis, J. G., & Manolakis, D. K. (2006). *Digital signal processing* (4th ed.). Prentice Hall.
37. *Pulsed Radiation.* (2021). Retrieved 25 March, 2021, from http://www.newport.com/Pulsed-Radiation/381844/1033/-content.aspx
38. Astaf'eva, N. M. (1996). Wavelet analysis: Basic theory and some applications Physics-Uspechi, 39(11), 1085–1108
39. *Vallen Systeme: The Acoustic Emission Company*. Retrieved 25 March, 2021, from https://www.vallen.de/downloads/
40. *SKOP-8M portable 8-channel acoustic emission device for non-destructive testing of materials, products and structure elements.* Retrieved 25 March, 2021, from https://www.ipm.lviv.ua/our_developments/our_developments.php
41. Bianchetti, R., Hamstad, M. A., & Mukherjee, A. K. (1976). Origin of burst-type acoustic emission in inflawed 7075–T6 aluminium. *Journal of Testing and Evaluation, 4*(5), 313–318.
42. Vinogradov, S. D. (1989). *Akusticheskii metod v issledovaniyakh po fizike zemletryasenii (Acoustic method in research into the physics of earthquakes)*. Nauka.
43. Gonorovskii, I. S. (1986). *Radiotekhnicheskiye cepi i signaly (Radio circuits and signals).* Radio and communication.
44. Stankevych, O., & Skalsky, V. (2017). The vibration of a half-space due to a buried mode I crack opening. *Wave Motion, 72*, 142–153.
45. Mushyk, E., & Myuller, P. (1990). *Metody prinyatiya tekhnicheskikh reshenii (Techniques for making technical decisions)*. Mir.
46. Smirnov, E. G. (1981). Akusticheskaya emissiya (Acoustic emission). *Nauka i technika (Science and Technology), 15*, 111–159.
47. Cousland, S. M., & Scala, C. M. (1981). Acoustic emission and microstructure in aluminium alloys 7075 and 7050. *Metal Science, 15*, 609–614.
48. Bibik, Z. I., & Nazik, V. D. (1984). Akusticheskaya emissiya pri plasticheskoi deformacii polikristallov aluminiya vysokoi chistoty (Acoustic emission during plastic deformation of high-purity aluminum polycrystals). *Metallofizika (Metal Physics), 4*(4), 92–99.
49. Harris, D. O., & Dunegan, H. L. (1974). Continuous monitoring of fatigue-crack growth by acoustic-emission techniques. *Experimental Mechanics, 14*(2), 71–80.
50. Mathis, K., & Chmelik, F. (2012). Exploring plastic deformation of metallic materials by the acoustic emission technique. In W. Sikorsky (Ed.), *Acoustic emission* (pp. 23–48). InTech.
51. Ivanov, Yu. F., Konovalov, S. V., Stolbushkina, O. A., et al. (2009). Evolyuciya poverchnosti razrusheniya alyuminiya, formiruyushcheisya pri polzuchesti materiala s nalozheniyem potencyala (Evolution of the fracture surface of aluminum formed during creep of material with potential imposition). *Fizika i chimiya obrabotki materialov (Physics and Chemistry of Material Processing), 5*, 80–83.
52. Osipov, V. G., Drobysheva, E. K., & Ushakov, E. V. (1964). Микромеханизмы разрушения (Micromechanisms of fracture). *Plasticheskaya deformaciya metallov (Plastic deformation of metals)* (pp 88–95).

53. Skal's'kyi, V. R., & Lyasota, I. M. (2010). Estimation of the heat-affected zone for the electron-beam welding of plates. *Materials Science, 46*(1), 115–123.
54. Klein, R. J. (Ed.). (2010). *Welding: Processes, quality, and applications. Mechanical engineering theory and applications.* Nova Science Pub Inc.
55. Skalskyi, V. R., & Koval, P. M. (2007). *Some methodological aspects of application of acoustic emission.* Publishing House Spolom.
56. Senthilkumar, S., Narayanan, S., & Denis Ashok, S. (2013). Acoustic emission–based monitoring approach for friction stir welding of aluminum alloy AA6063-T6 with different tool pin profiles. *Proceedings of the Institution of mechanical engineers Part B Journal of Engineering Manufacture, 227*, 407–416.
57. Rajaprakash, B. M., Suresha, C.N., Rachappa, & Upadhya, S. (2014). Application of acoustic emission technique for online monitoring of friction stir welding process during welding of AA6061-T6 aluminum alloy. In *TMS 2014 143rd Annual Meeting & Exhibition* (pp. 731–740). Springer.
58. Lee, Y.-W., Hossain Md. A. A. M., Hong, S.-T., Yum, Y.-J., & Park, K.-Y. (2011, June). *Characterization of friction stir spot welding of aluminum alloys using acoustic emissions.* Proceedings of the twenty-first International Offshore and Polar Engineering Conference, Maui, Hawaii, USA.
59. Venkitakrishnan, P. V., Sinha, P., & Krishnamurthy, R. (2006). Study and analysis of effect of various thermal processes in AA2219 annealed sheet using acoustic emissions. Materials & Design, 27, 770–775.
60. Gudramovych, V. S., Skalskyi, V. R., & Selivanov, Y. M. (2017). *Holografichne ta akustyko-emisiine diagnostuvannya neodnoridnych konstrukcii i materialiv (Holographic and acoustic emission diagnostics of inhomogeneous structures and materials).* Prostir-M, Lviv
61. Rabotnov, Yu. N. (1966). *Polzuchest elementov konstrukcii (Creep of structural elements).* Nauka.
62. Vittsel, V. I., & Edsit, N. R. (1977). Vliyaniye temperatury na razrusheniye (Influence of temperature on fracture). In *Razrusheniye (Fracture)* (Vol. 4, pp. 68–104). Mashynostroyeniye.
63. Skalskyi, V. R., Matviiv, Y. Y., & Krandinova, T. A. (2011). Metodyka ocinky kharakterystyk povzuchosti materialiv (Methods for assessing the creep characteristics of materials). *Mashynoznavstvo (Mechanical Engineering), 9–10*(171–172), 26–30.
64. Andreikiv, O. E., Skal's'kyi, V. R., Matviiv, Y. Y., & Kradinova, T. (2012). Evaluation of the durability of cracked plates under the conditions of long-term static loading and local creep. *Materials Science, 48*(1), 36–45.
65. Andreikiv, O. E., Skal's'kyi, V. R., Matviiv, Y. Y, & Dolins'ka, I. Y. (2013). Micromechanisms and a computational model of growth of low-temperature creep cracks in materials. *Materials Science, 49*(1), 25–35.
66. Stepanenko, V. O. (Ed.). (2002). *Materialoznavstvo i technologiya konstrukciinykh materialiv (Materials science and technology of structural materials).* Lybid.

Chapter 4
Evaluation of the Types and Mechanisms of Fracture of Composite Materials According to Energy Criteria

In recent years, the use of CM as a structural component in many branches of industry, including aerospace, rocket and space, chemical, automotive, for the production of pipes for power generating industry, mechanical engineering, civil engineering, and so on, is constantly increasing.

The structural integrity of composites depends on a few factors, including mechanical loads, defects that arise during their manufacture and operation, environmental conditions, and so on. Any of them can cause internal damage to CM. For long-term operation of products and structural elements made of such materials, it is important to consider the compatibility of composite components [1–6]. Thus, it is necessary to know the peculiarities of the fracture and its transition from the subcritical to the critical stage in the elaboration and operation of composite structures.

There are several stages of macrofracture in fiber CM: matrix cracking, delamination of fiber from the matrix, fiber fracture and elongation, displacement of matrix parts, and so on. To characterize the mechanisms of CM damage and fracture, different methods of NDT are used [7]; however, the AE method is preferred [8–10]. The AE signals are classified both by individual parameters (e.g., by the amplitude, frequency, time of increase of the leading edge of pulses, their duration, energy) and by their totality, in particular, by the method of pattern recognition. For time localization of the fracture features of the composites, the WT method is used in the analysis of AE signals.

This chapter presents the results of the investigation of the features of macrofracture of glass and aramid fiber-reinforced composites by the parameters of AE signals.

V. Skalskyi et al., *Acoustic Emission*, Foundations of Engineering Mechanics,
https://doi.org/10.1007/978-3-031-11291-1_4

4.1 Specific Features of Macrofracture of the Glass Fiber-Reinforced Composites

Because of the peculiarities of the structure and variety of constituent elements, considerable attention is paid to the study of the fiber-reinforced CM characteristics [11–14], the strength of which depends on the properties of the fibers. Fibers in a complex composite structure can redistribute stresses and block the propagation of small cracks that initiate in the matrix, thus making the abrupt brittle fracture almost impossible. In unidirectional composites, the anisotropy of properties (along and across the fibers) is observed. The resin matrix in the composite, which is distributed between the fibers, protects them, and keeps the fibers in the correct localization and orientation. The type of the matrix resin determines its absorption properties both to water (hygroscopicity) and to chemical compounds, mechanical properties at high temperatures, compressive strength, and mechanical rigidity.

Fiberglass is a common reinforcing element of CM. It is characterized by strength, heat resistance, high dielectric characteristics, low thermal conductivity, and abrasive strength. About 65% of a total number of CM based on fiberglass are produced today.

The research into the fracture of CM with fiberglass and various matrices are described in detail in the literature. In these studies, with the help of WT of AE signals the problems of location of AE sources, identification of their fracture mechanisms, determination of the residual life of structures with such materials, etc., are solved [15–23]. For this purpose, different types of WT are used: continuous, discrete, packet, with the analysis of the amplitude and frequency parameters, the energy of the levels of decomposition of AE signals.

The purpose of this research is to determine the dynamics of the development of fiberglass composite fracture using the energy criterion for identifying the types of fracture.

4.1.1 Substantiation of the Research Technique

Peculiarities of fiberglass composite fracture are studied using the criterion parameter E_{WT} of CWT and energy distribution of DWT for AE signals.

The choice of the wavelet for DWT and the depth of decomposition depend on the properties of a particular signal. There are no universal rules for such a choice. Many authors pay attention to the fact that for an effective choice it is necessary to study in advance the properties and characteristics of different wavelets, their features, as well as to have practical experience of work with a certain type of signals. However, some recommendations for choosing the type of wavelet, the number of levels of decomposition for certain problems of signal processing are known in the literature [24–26].

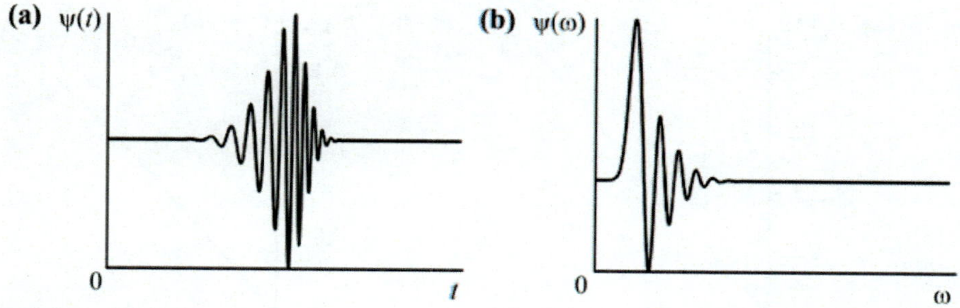

Fig. 4.1 Schematic representation of the wavelet function db20 (**a**) and its Fourier transform (**b**)

The analysis of the literature shows that DWT of AE signals is most often performed using Daubechies wavelets (db) [20, 27–33]. These functions are biorthogonal, with a compact carrier, rather regular, though asymmetric. Such properties are important for the analysis of nonstationary signals, which include AE signals. Other wavelets with similar properties, for example, symlets (sym) and coiflets (coif), can also be used with a slight difference in results.

For our research, the Daubechies db20 wavelet is chosen (Fig. 4.1).

This function is smooth; without sharp angles, it can easily be obtained numerically. Besides, it well approximates the AE signals registered during the composite materials fracture [7]. The spectrum of the db20 wavelet is closer to the band one than the lower-order wavelets. This provides less error in the frequency range [34]. Four levels of the decomposition are established. Separate software modules in Matlab7.1 are written for the DWT of AE signals.

Thus, using the criteria parameter E_{WT}, calculated according to CWT of AE signal, it is possible to establish the type of fracture currently occurring in the composite, and using the energy distribution of DWT—the correlation of composite fracture mechanisms and spectral characteristics of AE signals.

4.1.2 *Results and Discussion*

The fracture of fiberglass CM of hand-shaping and polyester resin (matrix material) under the quasi-static tension is investigated. Structure of the fiberglass composite consists of several layers of fiberglass, covered with a layer of resin with a hardener. The total thickness of fiberglass is 4 mm. The samples have a central crack with the final rounding radius commensurable with the smallest fiber radius ≈ 0.4 mm (see Fig. 3.37).

The experiments are performed on a tensile tear testing installation of SVR-5 type with a loading speed of 4×10^{-7} m/s. The Kaiser effect is used to avoid the impact of the signals from the installation clamps on the research results. The samples shunted in a special mandrel being preliminary loaded. In order to prevent the influence of power supply interference, the method of parallel (indicator) channel is applied.

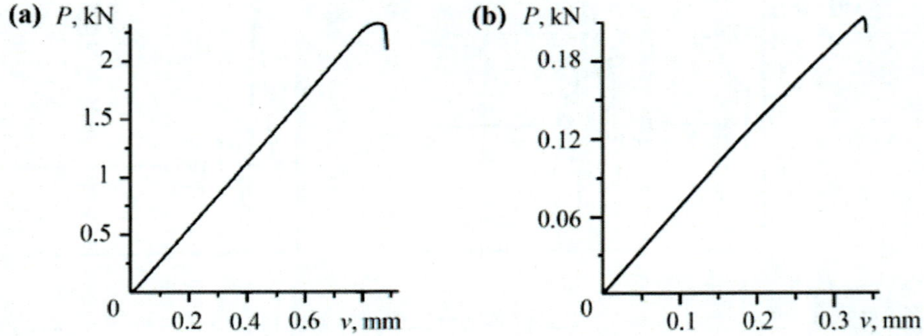

Fig. 4.2 Fracture curves of the samples with a central crack in the "load *P*—crack tip opening displacement *v*" coordinates: **a** is a hand-shaping composite with an arbitrary orientation of fiberglass fibers in a polyester matrix; **b** a sample of polyester resin

The location of the sensors, the parameters of the system of selection, amplification, and processing of AE signals for all the samples are the same. AE information is detected by the SKOP-8 M system [35] with two measuring channels, one of which is indicator. The selection modes are as follows: sampling duration 0.5 ms; period of discretization of the analog signal is 0.25 μs; cutoff frequency low-pass filter is 600 kHz, high-pass filter is 200 kHz; discrimination threshold is within 30%. The AE gain factor is 70 dB (40 dB for the preamplifier). The frequency band of the measuring channel is 0.2–2.0 MHz.

Since the discretization step of the analog signal is 0.25 μs, according to the DWT algorithm, the highest frequency in the AE signal will be equal to 2000 kHz. Then we get the following correspondence of the decomposition levels of the DWT and frequency bands: D1—1000–2000 kHz; D2—500–1000 kHz; D3—250–500 kHz; D4—125–250 kHz; A4—0–125 kHz. The level of D1 (1000–2000 kHz) is not analyzed because it is beyond the limits of the measuring channel frequencies.

Fracture curves are constructed in the post-operation mode (Fig. 4.2), and the CWT and DWT of the detected AE signals are analyzed.

It is found (Fig. 4.2) that the fracture of the composite and matrix material is quasi-brittle, which is typical for the fiber composites. According to the SIF values of the macrofracture start K_{IAE}, calculated by the AE method, the matrix (K_{IAE} = 0.31 MPa) is more brittle than the fiberglass (K_{IAE} = 3.36 MPa).

Already in the early stages of fracture, the measuring system has detected AE signals, the characteristic appearance of which is illustrated in Fig. 4.3a.

Initially, they have small amplitudes, which later increase, and according to the spectral analysis, their dominant frequencies are in the range of 200–250 kHz (Fig. 4.3b). The number of such signals is 90% of all registered.

According to the DWT (Fig. 4.3c), the maximum energy for these signals is concentrated at the level of D4 decomposition (frequency range is 125–250 kHz). Let us analyze the known literature data. In [36], during fracture of a carbon fiber composite, the relationship between the fracture mechanisms and the maximum frequencies of AE signals is established: For matrix cracking, the maximum

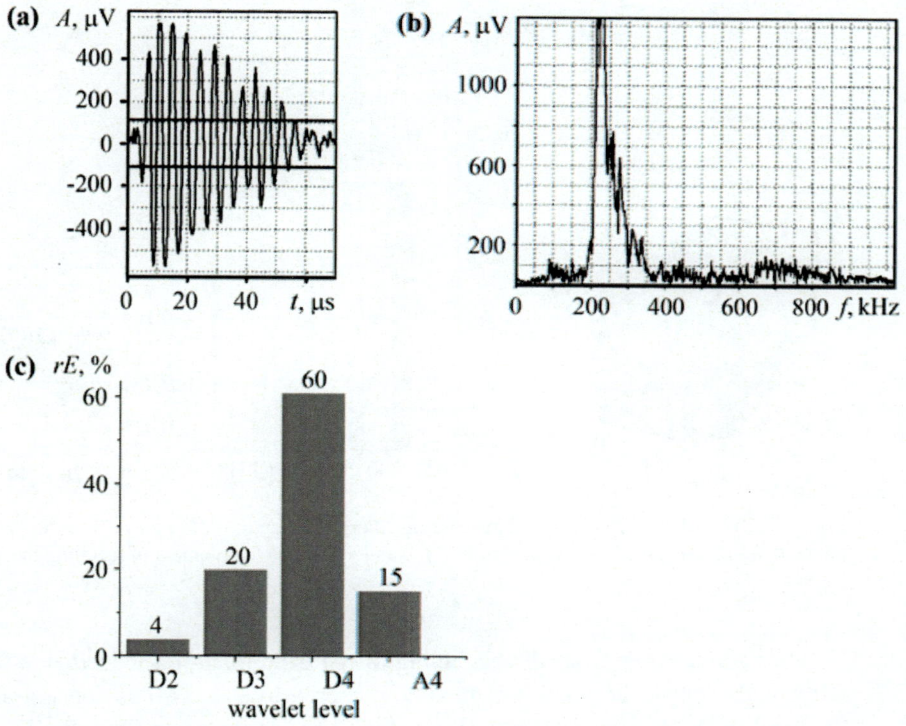

Fig. 4.3 Wave (**a**) and amplitude–frequency (**b**) representation of a typical AE signal registered during fracture of the fiberglass composite; the energy distribution of AE signals according to the DWT (**c**)

frequency is < 100 kHz; for fiber/matrix delamination the maximum frequency is in the range of 200–300 kHz; for carbon fiber rupture – in the range of 400–450 kHz. Similar regularities are obtained for fiberglass composites in [20, 21]. Dominant frequency range of AE signals accompanying matrix cracking is in the range of 50–150 kHz, fiber/matrix delamination—in the range of 170–350 kHz; fiber fracture—in the range of 300–400 kHz.

Figure 4.4 shows the CWT of the typical AE signal under the fiberglass composite fracture. It is found that with increasing the signal amplitude, the maximum wavelet coefficients WT_{max} increase. A characteristic feature of their CWT is a narrow frequency band in the WT–f coordinates (Fig. 4.4b) and also a significant duration of radiation when reaching the value of WT_{max} (Fig. 4.4c).

The criteria parameter E_{WT}, calculated by relation (3.3), is in the range corresponding to the brittle type of fracture. The duration Δt of these signals is in the range of 10–20 μs, and the bandwidth Δf is in the range of 65–75 kHz. Initially, microcracks are formed at the fiberglass/matrix boundary, which is accompanied by the AE signals with an energy parameter $0.021 \leq E_{WT} \leq 0.1$. Then, under the action of load, the microcracks join to form macrocracks, which is accompanied by AE signals with parameter $0.1 < E_{WT} \leq 0.265$.

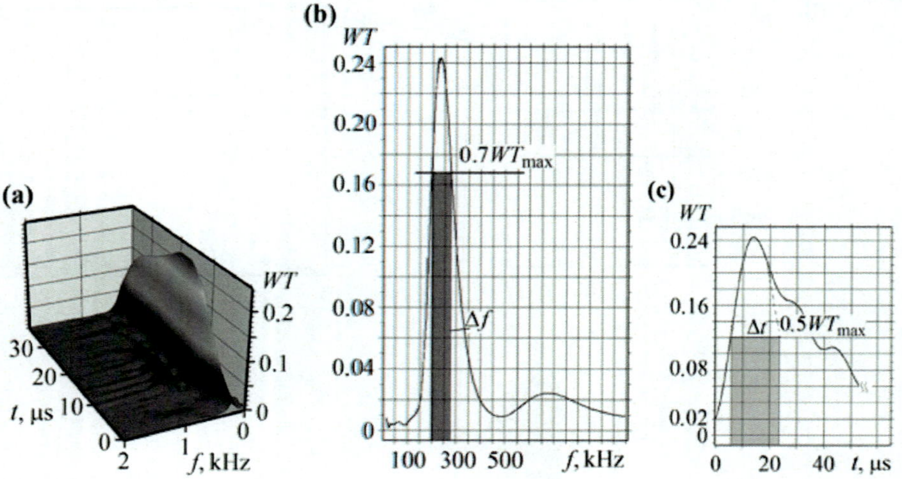

Fig. 4.4 CWT of the typical AE signal during fiberglass fracture: 3D plot of wavelet spectrum intensity **a** 2D projections on the planes WT–f **b** and WT–t **c** at the moment of reaching the WT_{max} value

After about 400 s from the beginning of the experiment (at load $P = 0.6$ kN), AE signals appear, in which the maximum energy by the DWT is concentrated at the level of A4 wavelet decomposition (Fig. 4.5). At the same time, the D4 level (125–250 kHz) dominates among the detailing levels. Since the lowest frequencies (0–125 kHz) correspond to the A4 level for the DWT, it can be assumed that the matrix cracking occurs simultaneously with the delamination.

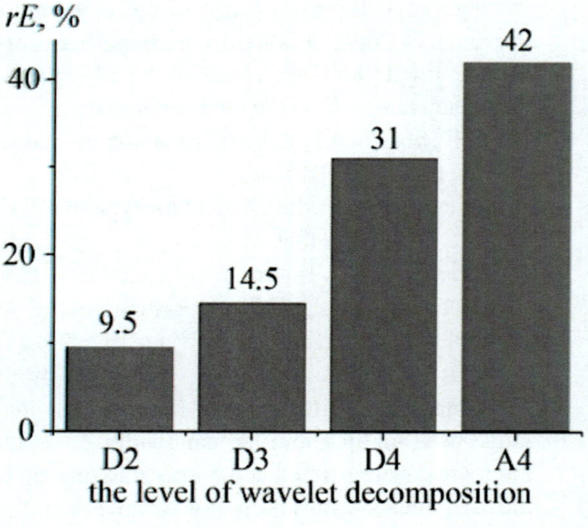

Fig. 4.5 Energy distribution of AE signals generated after reaching the load $P = 0.6$ kN

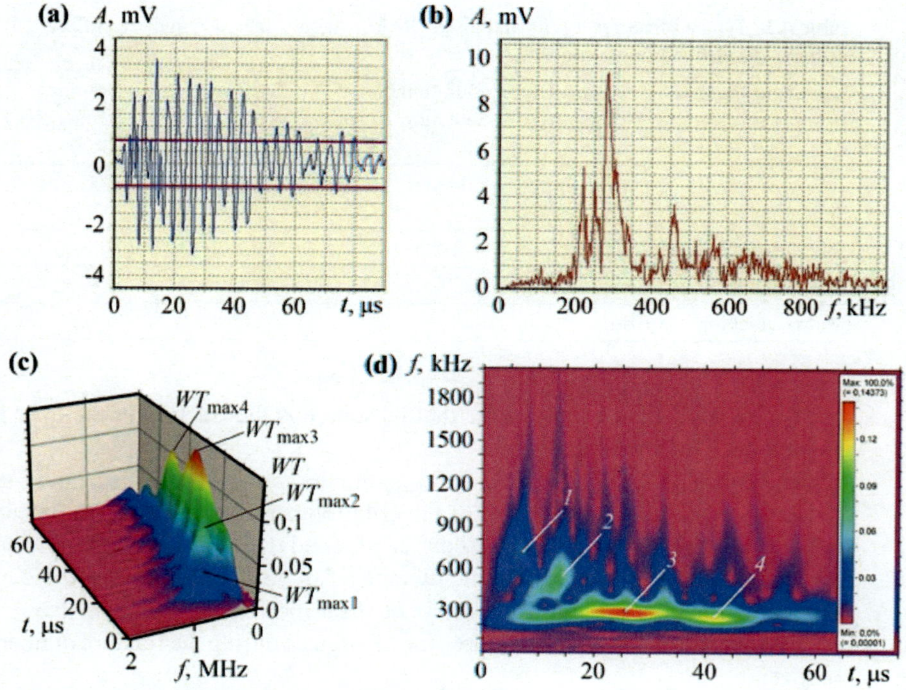

Fig. 4.6 Typical AE signal at the supercritical stage of composite fracture: **a** waveform, **b** frequency spectra, **c** 3D-representation of CWT, **d** 2D-time projection

Further, AE signals with several local CWT maxima at different frequencies are detected (Fig. 4.6), which may indicate the almost simultaneous development of different mechanisms of composite fracture.

For example, in the signal shown in Fig. 4.6c, d, four local maxima are identified: *1*—at time $t = 6.88$ μs from the beginning of the AE event, for which $\text{WT}_{\text{max1}} = 0.043$ and $f_{\text{max1}} = 720$ kHz; *2*—$t = 13.38$ μs, $\text{WT}_{\text{max2}} = 0.08$, $f_{\text{max2}} = 520$ kHz; *3*—$t = 26.13$ μs, $\text{WT}_{\text{max3}} = 0.14$, $f_{\text{max3}} = 300$ kHz; *4*—$t = 41.38$, $\text{WT}_{\text{max4}} = 0.12$, $f_{\text{max4}} = 250$ kHz.

According to the DWT, the maximum energy of the studied signal is concentrated at the D3 level, which corresponds to the frequency range of 250–500 kHz. This range includes the frequencies f_{max3} and f_{max4} of the local maxima WT_{max3} and WT_{max4} of this AE event (Fig. 4.6c).The duration of these pulses is $\Delta t_3 = 22$ μs and $\Delta t_4 = 14$ μs (Fig. 4.6d), and the value of the energy parameter is $E_{\text{WT3}} = 0.389$ and $E_{\text{WT4}} = 0.142$, respectively. Since the CWT parameters with WT_{max4} event coincide with those described above (see Figs. 4.4 and 4.6), it can be assumed that its source is the fiber/matrix debonding or the propagation of macrocracks along the fiber/matrix boundary.

For the local event WT_{max3}, the duration and energy parameter are greater than for typical AE signals. Since $f_{\text{max3}} = 300$ kHz, then taking into account the known

Table 4.1 Typical features of local pulses of AE signals that accompany different fracture mechanisms of fiberglass composite

Dominant fracture mechanism	Type of fracture	Duration of radiation Δt (μs,)	Bandwidth Δf (kHz)	Dominant frequencies f_{max} (kHz)
Fiberglass/matrix debonding	Ductile–brittle Brittle	10–20	65–75	125–500
Fiber pullout	Ductile	4–10	230–270	500–700
Fiber rupture	Brittle	>20	80–90	400–500
Matrix cracking	Brittle	–	–	<125

results [26, 31, 37], we can assume that its source is the fracture of the fiber in the glass cloth.

For local maxima WT_{max1} and WT_{max2}, the duration of events is small—$\Delta t_1 = 5$ and $\Delta t_2 = 6.5$ μs, and the value of the criterion parameter is $E_{WT1} = 0.0068$ and $E_{WT2} = 0.026$. This corresponds to the ductile and ductile–brittle type of fracture, respectively. Taking into account the literature data [37], the source of these events is probably the fiber pullout from matrix or shear mechanisms in the matrix.

Table 4.1 shows the typical features for AE signals during the fracture of fiberglass composite.

Thus, using the time–frequency analysis and the energy criterion for identifying types of material fracture, it is possible to establish the relationship between the parameters of CWT and DWT of AE signals and the mechanisms of fracture of fiberglass composite (fiber/matrix debonding, fiber rupture, matrix cracking, fiber pullout from the matrix or shear mechanisms in the matrix).

4.2 AE-Diagnostics of Fracture of the Aramid Fiber-Reinforced Composites

Recently, the polymer, in particular aramid, fibers [38], which first became known as Kevlar, have been widely used in CM.

4.2.1 General Characteristic of the Aramid Fibers

The aramid fibers belong to the class of aromatic polyamides. Their chemical structure is characterized by a high degree of orientation and rigidity. These characteristics, in particular, are caused by a large number of aromatic (benzene) rings. The structure of the aramid fiber is referred to cross-linked polymers. Rigid polymer chains are in a straightened position and form a very dense package in the fiber volume, which determines high mechanical characteristics of such fibers. The crystalline nature of the polymer provides thermal resistance of the fibers, and aromatic rings in the structure of the macromolecule supply their chemical resistance. Having a rigid cross-linked structure of macromolecules, the aramid fibers do not undergo any phase transformations during heating up to the temperature of thermal decomposition.

Among organic polymer fibers, the aramid fibers have the highest operational characteristics (strength in the range of 3.6–3.8 GPa, modulus of elasticity in the range of 62–180 GPa). The most common of these are para-aramid (Kevlar, Twaron) and meta-aramid (Nomex) fibers. They are prepared in the form of technical yarns with different linear densities and structures. The number of elementary fibers in typical yarns can vary from 130 to 1000 for fabric production and from 500 to 10,000 for ropes and cord. The fibers practically do not show creep under load. In the whole loading range up to the fracture, the dependence of stresses on deformations is linear.

The properties of the aramid fibers are significantly affected by the structure [39], microscopic deformations [40], as well as twisting of the yarns, which leads to the surface damage. As a result, as the degree of twisting increases, the modulus of elasticity and strength of the fibers decrease markedly. This effect is thought to be related to their surface damage during twisting. This assumption is confirmed by the results of fatigue tests of fibers, which show that the fibers can withstand a large number of load cycles, if they are not subjected to surface friction, and their strength under friction is very low. Therefore, for the production of appropriate composites, it is important to provide the greatest protection of the surface of the fibers and their adhesion to the matrix [41]. For this purpose one of the following methods is used: finishing (surface impregnation), plasma treatment, ion etching, and so on. Recently, plasma treatment methods have become especially important for the modification of fibrous materials, which consist in the influence of gas discharges plasma on the material (smoldering, barrier, corona, spark, arc, high-frequency, or ultrahigh-frequency discharges) [42]. They make it possible to purposefully change the structure of the fiber polymer in order to obtain the necessary physicomechanical, surface, and operational properties of the fibers. High-frequency discharges are increasingly used to modify synthetic fibrous materials. A small percentage of the ultraviolet component of the plasma of high-frequency capacitive discharge provides modification of materials without their fracture and with a stable plasma effect. In particular, by activating the surface after plasma treatment, a significant improvement in the adhesive ability of the aramid fibers and their strength characteristics is achieved [43–45].

Aramid fibers are widely used in industry. Combining lightness and strength, aramids are indispensable during the design of aerospace equipment, car parts, shipbuilding, production of protective equipment (chemical, fire, military), etc.

4.2.2 Influence of Various Factors on the Mechanical Properties of CM Reinforced with the Aramid Fibers

The influence of the binder on the mechanical properties of the composite is determined by its two parameters: adhesion and modulus of elasticity [8, 12, 13], which affect the CM type and, finally, characterize the strength characteristics of the composite. Low modulus binders usually cannot provide the load redistribution between the reinforcing fibers throughout the composite volume. However, the individual fibers are deformed independently of each other. The fracture (rupture) of one fiber causes a significant redistribution of load in its vicinity and overstrain of the neighboring fibers. As a result, there is an avalanche-type fracture of the material. The average fiber strength in the composite is low. If the selected combination of adhesion and modulus of elasticity provides approximately uniform load distribution between all fibers, the average (effective) strength of the composite will be higher. In practice, it is possible to get some average value of the potential strength of the material. As the volume content of the fiber increases, the strength of the material usually increases at first, and when a certain degree of filling is reached, it stabilizes or decreases. For example, an increase in temperature leads to a certain decrease in the strength of microplastics made of the aramid fibers, and its decrease has almost no effect on the strength properties.

The dependence of stresses on the deformations of the composites based on the aramid fibers is close to linear. Action of a constant load for a long time on the samples made of unidirectional composite based on the aramid fibers is accompanied by an increase in their deformation—creep. As the load increases, the deformation caused by creep increases, and with its long-term action, the dependence of the deformation on time becomes linear for a wide range of initial loads. The rigidity and strength of CM based on the aramid fibers under load, transverse relative to the direction of reinforcement, is much less than for the load in the direction of reinforcement.

4.2.3 Mechanisms of Unidirectional CM Fracture

A number of models for determining the strength of composites with unidirectional fibers are known in the literature [8, 12, 13]. High-strength and rigid fibers, which provide high characteristics of CM, fracture mostly brittle. They contain arbitrarily arranged defects and cracks, which is typical for the brittle materials. Some of them are necessarily fractured by very low tensile forces.

According to Rosen's statistical model [13], with increasing load, there is a statistical accumulation of fiber breaks in the composite, until a sufficient amount of damage occurs in some cross section of the material, which will reduce the strength and lead to failure.

According to Parrat's model [12], fracture occurs when, due to the accumulation of breaks, the length of the fibers decreases so much that a further increase in load cannot be transmitted to them due to the achieved shear strength of the matrix. Thus, breaks of the fibers eventually cause shear fracture of the matrix and, consequently, the composite fracture. The model of fracture because of damage accumulation gives the upper limit of the fiber composite strength, which cannot be achieved due to the emergence of additional types of fracture.

According to these models, the fracture of a unidirectional fiber composite during monotonically increasing load occurs gradually by the accumulation of the yarn breaks. In this case, the transverse cracks of the fibers do not propagate into the matrix. This is explained by the large-scale effect of fiber strength: a shorter fiber is stronger (on average). The maximum load corresponds to the breaks of all fibers into individual links of critical length: If the length of the fiber is less than the critical one, it is removed from the matrix without breaking.

Therefore, the following possible mechanisms of fracture of unidirectional fiber composites during tension are considered: crack propagation across the fibers; propagation of the crack by the boundary between the fiber and the matrix along the entire length of the sample; accumulation of fiber breaks in the sample with increasing load, which eventually leads to a weakening of the cross section of the sample and the fracture of the composite.

To study the mechanisms of the fracture of composites reinforced with the aramid fibers, Morgan and co-authors [40] propose a model of finite-chain distribution of macromolecules in a fiber (Fig. 4.7). To do this, a number of fractographic studies are performed, by studying the structure of crystallites; the procedure of fiber production and the physical processes that accompany it; microscopic deformations and mechanisms of fracture of the composite with the aramid fiber and the fibers themselves; topography of fiber fracture by different methods of their processing. According to this model, macromolecules are aligned parallel to the axis of the fibers. In the outer part of the fiber, the chain of macromolecules is located randomly relative to the others, but in the inner fiber it ends in a cluster.

Taking such model of the aramid fiber structure into account, the authors consider the corresponding scheme of crack propagation through macromolecular finite chains in the fiber shell and core (Fig. 4.8a), which is confirmed by the electronic microfractogram of Kevlar 49 fiber fracture topography (Fig. 4.8b).

When investigating the fracture of composites reinforced with the aramid fibers, the AE method has been effectively used [46–58], for example, to determine the

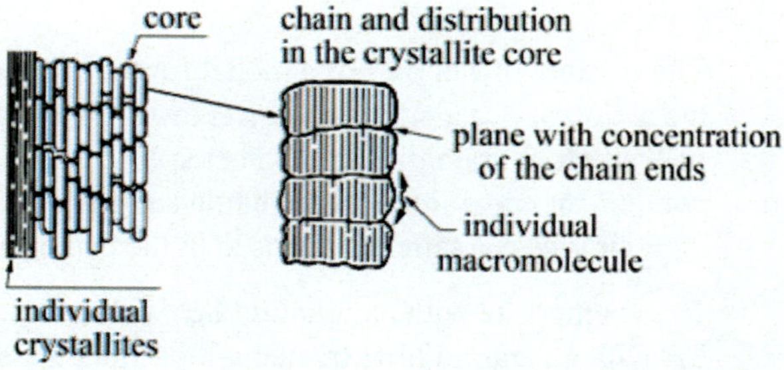

Fig. 4.7 Model of finite-chain distribution of macromolecules in aramid fiber

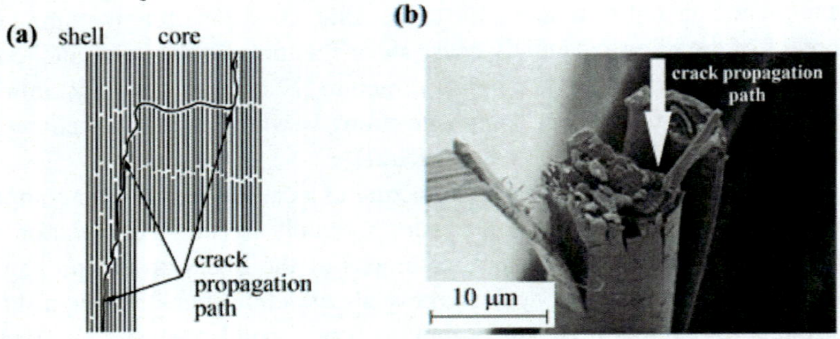

Fig. 4.8 Path of crack propagation through macromolecular finite chains in the shell and core of the aramid fiber **a** and the topography of Kevlar 49 **b** fiber fracture

basic mechanical parameters of fibers, features of creep and fatigue failure [47], to study the types of damage and fracture mechanisms [48–57], to determine the shear stresses at the fiber/matrix boundary [58], etc. The parameters of AE signals, such as amplitude or amplitude distribution, counting speed, and total energy, are most often analyzed. In [56, 57], a correlation between the fracture mechanisms of composites with the aramid fibers and the dominant frequencies of AE signals is established by means of the window Fourier transform. However, information on the application of WT for the analysis of AE signals during fracture of such composites is unknown to us.

4.2.4 Substantiation of the Research Methods

Peculiarities of the fracture of aramid fiber-reinforced composite are studied by the criteria parameter E_{WT} of the CWT (Eq. 3.3) and energy distribution of the DWT of AE signals. Daubechies wavelet db20 and four levels of decomposition are used to decompose the AE signal by the DWT.

4.2.5 Results and Discussion

The fracture of samples of a model for the aramid fiber-reinforced composite under the action of quasi-static tension is investigated. The composite structurally consists of a single p-aramid fiber (diameter 10 μm) of the grade Twaron 1014 in a two-component epoxy matrix (the hardener/ epoxy rubber ratio is 50:50).

Four types of samples are used in the experiments:

1. a composite with an aramid fiber without treatment (test sample);
2. with an aramid fiber treated with a silicone oil;

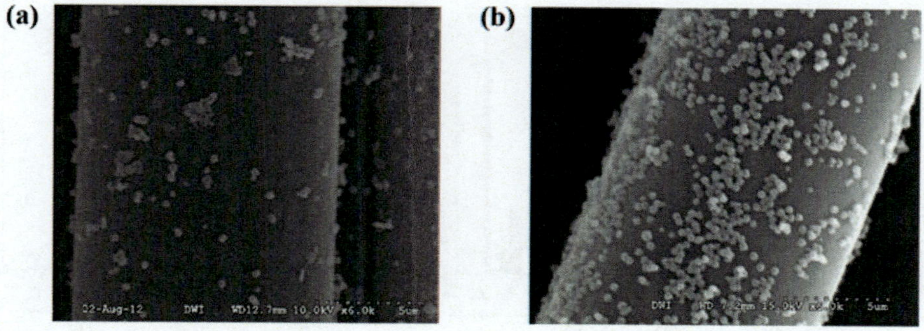

Fig. 4.9 Fibers of type 3 (**a**) and 4 (**b**) treated with polystyrene methacrylate nanoparticles

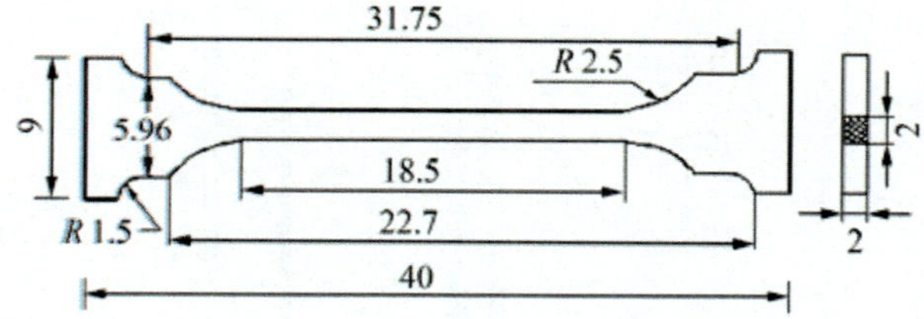

Fig. 4.10 Geometry of the composite sample with an aramid fiber (mm)

3. with an aramid fiber treated in plasma medium and in polyethyleneimine with the addition of 10% poly(styrene-co-glycidyl methacrylate) nanoparticles (Fig. 4.9a);
4. with an aramid fiber treated in plasma and polyethyleneimine with the addition of 100% poly(styrene-co-glycidyl methacrylate) nanoparticles (Fig. 4.9b).

Samples (Fig. 4.10) are tensile tested on a rupture machine SVR-5 with a loading speed of 4×10^{-7} m/s. The experimental conditions and settings of the measuring equipment are the same as for the fiberglass composite.

In the post-operation mode, the stress–strain curves are constructed and the CWT and DWT of AE signals are analyzed according to the methods proposed above. All the data are statistically analyzed using the ANOVA package.

Figure 4.11 shows the typical stress–strain curves and the corresponding AE activity curves during fracture of the CM with aramid fiber using different processing methods. Table 4.2 presents the average values and standard deviations of the stresses of the initial σ_{AE} and final σ_f fracture.

It is found (Fig. 4.11) that the fracture of the samples of composites of types 1 and 2 is quasi-brittle, and of types 3 and 4—elastic–plastic. This change in the nature of fracture can obviously be a result of the treatment of the aramid fiber in plasma and polyethyleneimine with nanoparticles. This reinforces the adhesion between the fiber and the matrix and thus increases fracture resistance during tension of the composite as a whole.

It is found (Table 4.2) that the impregnation of the aramid fiber with a silicone oil strengthens of the composite, the loading of which at the beginning of fracture

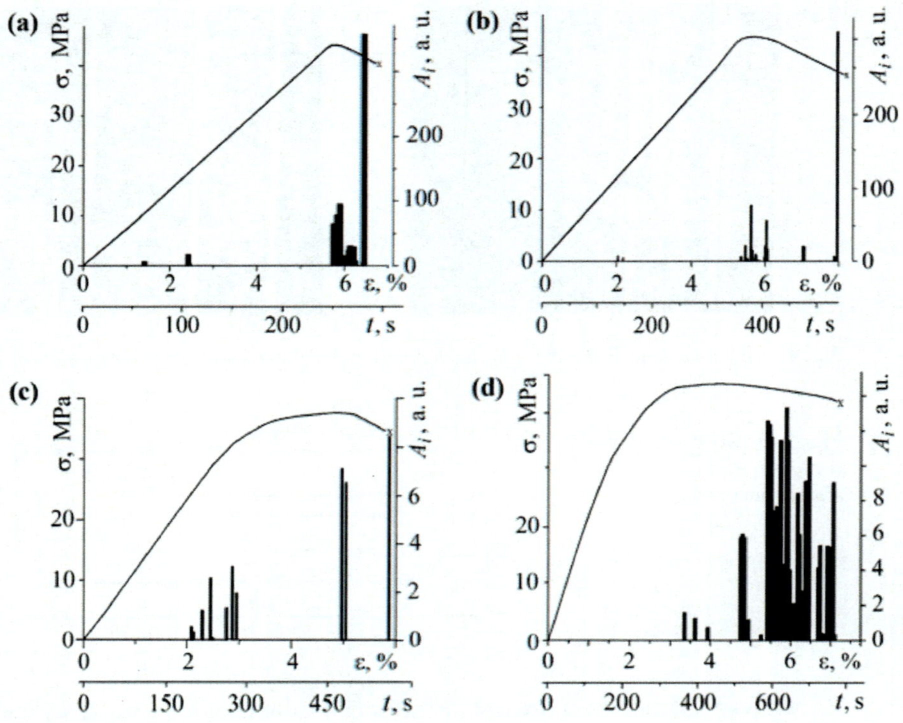

Fig. 4.11 Typical stress–strain curves and distribution of the amplitudes of AE signals by samples during fracture of CM with aramid fiber of different processing methods: **a** composite type 1; **b** 2; **c** 3; **d** 4

Table 4.2 Average values and standard deviations (std) of the initial fracture stress σ_{AE} and final fracture stress σ_f under tensile loading of composite samples reinforced with an aramid fiber using different processing methods

Composite type	Initial fracture stresses σ_{AE}: (±std) (MPA)	Final fracture stress σ_f: (±std) (MPA)
1	12 (±3.8)	37.6 (±3.5)
2	20 (±2.3)	39.5 (±2.7)
3	25 (±1.6)	39.7 (±2.19)
4	40 (±2.5)	39.7 (±2.36)

increases by 66% (12 vs. 20 MPa), and the final fracture—by 5% (39.5 vs. 37.6 MPa), compared with the composite with untreated fiber. The following factors indicate a reinforcement of the adhesion between the fiber and the matrix in the composite after fiber treatment with a silicone oil: (1) differences in the generation of AE – in the composite of type 2 (Fig. 4.11a) AE begins to be generated at the greater deformation than of type 1 (Fig. 4.11b) (2.2 vs. 1.8%); (2) loading time prior to failure: composite of type 1 ruptures in 350 s ($\varepsilon = 6.5\%$) from the beginning of the experiment, and of type 2 – in 550 s ($\varepsilon = 7\%$).

Due to the aramid fiber treatment in plasma and polyethyleneimine with nanoparticles, the load of the beginning of fracture increases significantly (Table 4.2): compared to the composite with untreated fiber—twice for composite of type 3 and three times for the composite of type 4. Also, the complete failure load increases by almost 6% (37.6 MPa for the composite of type 1 and 39.7 MPa for types 3 and 4). Although the load of complete fracture of the composites 3 and 4 is almost the same, the time to achieve it is different—580 s ($\varepsilon = 6\%$) and 780 s ($\varepsilon = 7\%$), respectively (Fig. 4.11c, d). In the sample of type 3, AE begins to be generated at deformation equal to 2–3% and increases at deformation of 5–6%. In the sample of type 4, AE begins to be generated at deformation equal to 3–4% and increases at deformation equal to 5–7%. It can be concluded that the increase in the number of nanoparticles during fiber processing promotes the interaction and formation of new bonds between its surface and macromolecules of the matrix. This increases the strength and ductility of the composite. The AE analysis during tension of composites shows that after the aramid fiber treatment in plasma and polyethyleneimine with nanoparticles, the sum of amplitudes of AE signals significantly decreases, which also indicates a decrease in the fracture brittleness of the composite.

Therefore, the treatment of the aramid fiber significantly affects the strength of the reinforced composite. Both impregnation with a silicone oil and treatment in plasma and polyethyleneimine with nanoparticles contribute to the increase in its strength characteristics. The composite of type 4 with the aramid fiber treated in plasma and polyethyleneimine with the addition of 100% polystyrene methacrylate nanoparticles shows the greatest fracture resistance.

4.2.6 Analysis of AE Signals During Tension of the Composite of Type 1

Figure 4.12 shows a typical AE signal during fracture of the aramid fiber-reinforced composite of type 1. AE signals with larger and smaller amplitudes are generated. Two types of signals are distinguished according to the spectral analysis: I—with dominant frequencies in the range of 200–300 kHz (65% of the total number of registered); II—in the range of 500–600 kHz (35% of the total number of registered). Figure 4.13 shows the energy distribution according to the levels of the DWT decomposition, which is consistent with the results of spectral analysis.

Figure 4.14 shows the time distribution of AE signals registered in the experiment by the energy parameter, dominant frequency, and duration of radiation.

In the initial stages for AE signals, the energy parameter is $0.017 \leq E_{WT} \leq 0.148$, the duration Δt is in the range of 15–25 μs, and the frequency f_{max} corresponding to the wavelet coefficient WT_{max} is in the range of 200–300 kHz. According to the DWT, such signals are classified as Type I, and according to the energy criterion, they characterize the ductile–brittle and brittle fracture.

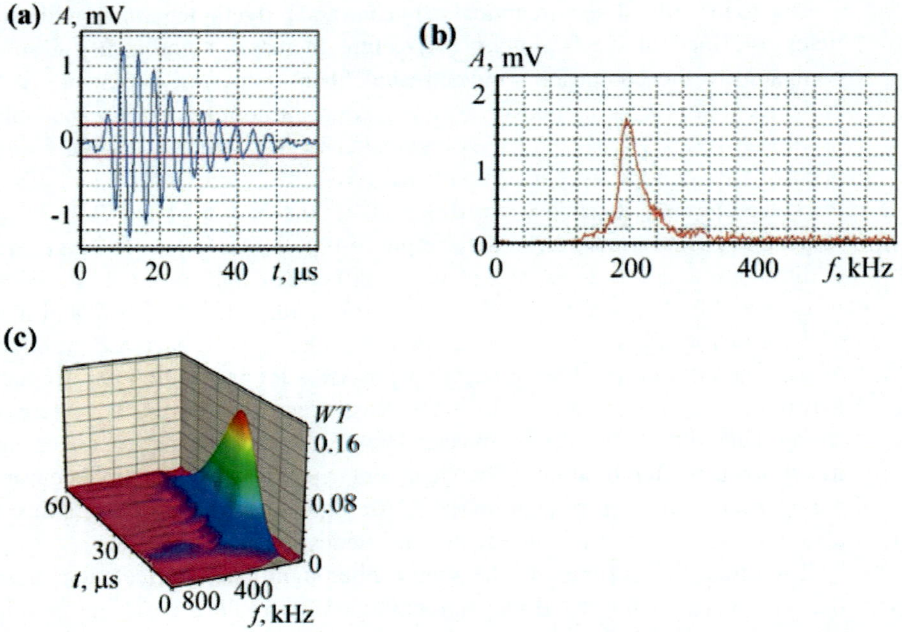

Fig. 4.12 Waveform (**a**), spectral characteristics (**b**), and CWT (**c**) of the typical AE signals during tensile test of the composite of type 1

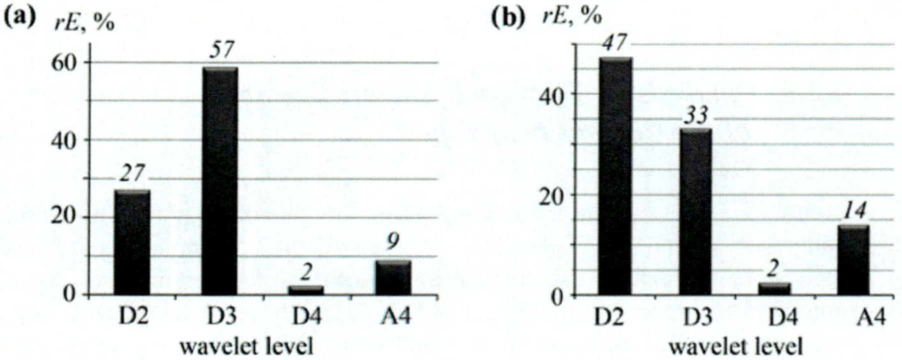

Fig. 4.13 Diagrams of energy distribution by the levels of the DWT decomposition of two types of AE signals during tension of the composite of type 1: **a** Type I; **b** Type II

In the time interval of 275–325 s from the beginning of the experiment, when the deformation of 5.5% is achieved (Fig. 4.11a), there is an active development of fracture in the composite. This fracture is accompanied by many AE signals. At this time (Fig. 4.14a) AE signals of Type I, which according to the energy criterion correspond to the ductile–brittle and brittle fracture, and of Type II, which characterize ductile and ductile–brittle fracture, alternate. The range of the energy parameter E_{WT} is

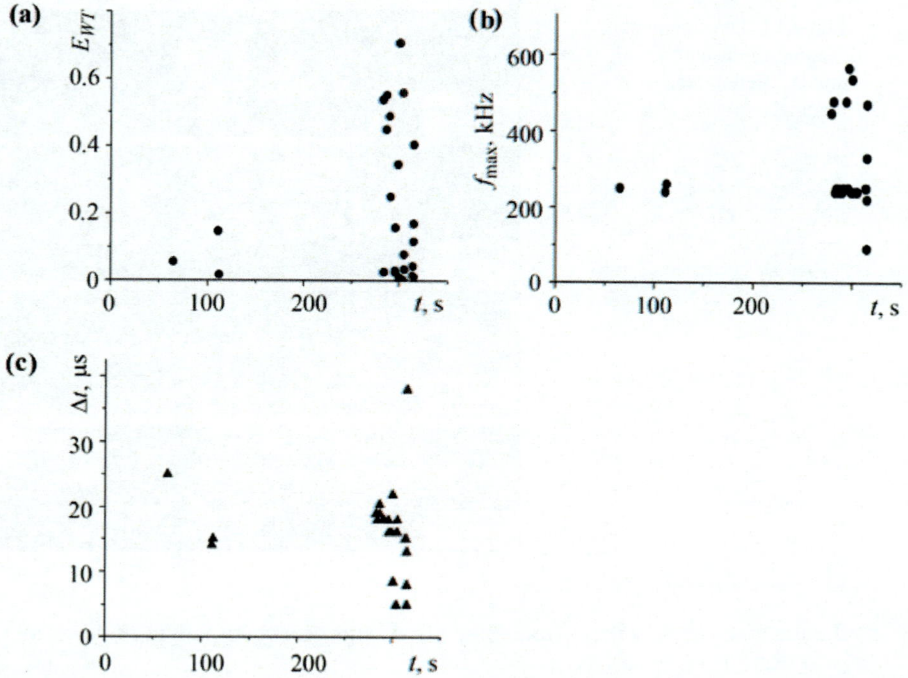

Fig. 4.14 Time distribution of AE signals by energy parameter (**a**), dominant frequency (**b**), and duration of radiation (**c**) during tension of the composite of type 1

$0.118 \leq E_{WT} \leq 0.708$ for brittle fracture, $0.011 \leq E_{WT} \leq 0.079$ for ductile–brittle, and $0.00219 \leq E_{WT} \leq 0.00846$ for ductile rupture.

Figure 4.15 illustrates the development of fracture in the composite with the aramid fiber in tension. The results have been obtained by photopolarization method at the Leibniz Institute of Interactive Materials (Aachen, Germany). As we can see, as the load increases, the fiber intensively delaminates from the matrix: at first, single sites of delamination appear. Such defects merge and spread along the fiber with increasing load. Therefore, we can assume that the AE signals classified as Type I accompany this process. The initial stages are characterized by AE signals with the energy parameter $0.017 \leq E_{WT} \leq 0.056$, and at the stage of intensive development of fracture, the signals with the parameters $0.011 \leq E_{WT} \leq 0.079$ and $0.118 \leq E_{WT} \leq 0.708$ alternate.

Considering the results of [40], we can assume that the fiber fracture (propagation of longitudinal and transverse microcracks in the fiber) (see Fig. 4.8) is the source of AE signals of Type II with the energy parameter $0.011 \leq E_{WT} \leq 0.18$. The fracture is of the ductile–brittle and brittle type in this case.

The obtained results are in agreement with the known ones [55], where the analysis of polarization microphotography and AE activity shows that during model composite tension the aramid fiber breaks locally in the places of delamination from

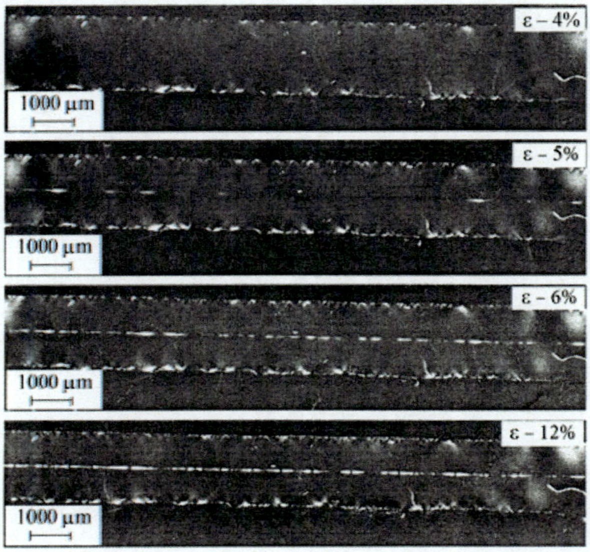

Fig. 4.15 Development of fracture in the composite under different tensile loading

the matrix. The dominant frequency of AE signals accompanying this break [56] is over 300 kHz (Fig. 4.16).

For the event with the wavelet coefficient WT_{max2}, the energy parameter $E_{WT_{max2}} = 0.406$ and the dominant frequency $f_{WT2} = 220$ kHz. This event corresponds to the brittle fracture. According to the DWT, the event is of Type I, which characterizes the propagation of delamination between the fiber and the matrix.

Local events are also recorded at frequencies below 100 kHz, and their energy parameter E_{WT} corresponds to brittle fracture. According to the known literature data [38, 56, 57], they are generated due to the crack propagation through the cross section of the sample, which eventually leads to its complete destruction. The events with an energy parameter $0.00219 \leq E_{WT} \leq 0.00846$ are generated, probably, taking to

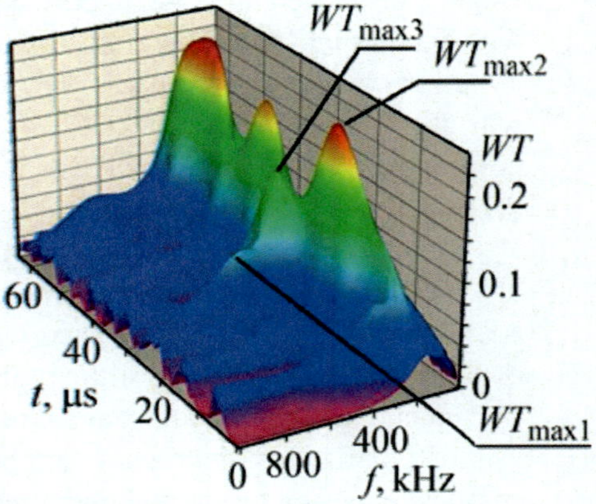

Fig. 4.16 CWT of AE signal at the final stage of fracture of the composite of type 1

account of the results of [48], during the development of diagonal shear of the matrix parts before dividing the sample into two pieces.

4.2.7 Peculiarities of AE Signals During Tension of the Composite of Type 2

Figure 4.17 illustrates the time distribution of the AE signals detected during the experiment according to the energy parameter, dominant frequency, and duration of radiation.

Contrary to the previous case, during tension of this composite, AE signals are generated in three frequency bands (Fig. 4.17b): 220–250 kHz (75% of the total number of registered); 300–400 kHz (6%); 500–600 kHz (9%).

In the initial stages of fracture, AE signals with an energy parameter $0.00765 \le E_{WT} \le 0.015$ and a dominant frequency range of 220–250 kHz accompany the delamination of the fiber from the matrix. This delamination occurs in ductile and ductile–brittle manner with the parameter $0.00028 \le E_{WT} \le 0.0005$ and the dominant frequency range of 500–600 kHz. So we can conclude that the crack propagates ductile in the fiber itself. This change in the fracture character of the composite of

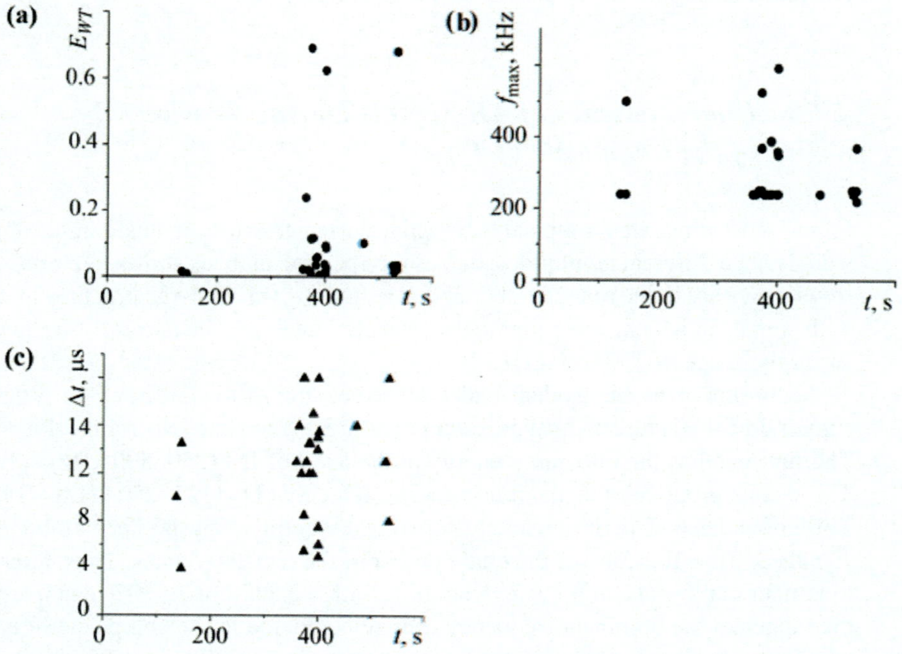

Fig. 4.17 Time distribution of AE signals by energy parameter (**a**), dominant frequency (**b**), and duration of radiation (**c**) during tension of the composite of type 2

type 2 compared to the composite of type 1 is apparently caused by the treatment of the fiber with a silicone oil. With an increase in the load for the deformation of 5.3% in the range of 350–420 s from the beginning of the experiment (later than in the previous case), the fracture of the composite intensifies (see Fig. 4.11b), which is accompanied by the generation of a large number AE signals with different energy parameters and dominant frequencies.

The alternation of AE signals is observed that accompany the delamination of the fiber from the matrix (frequency range of 220–250 kHz; energy parameter range $0.013 \leq E_{WT} \leq 0.087$ for ductile–brittle fracture, for brittle it is $0.102 \leq E_{WT} \leq 0.691$), propagation of longitudinal (frequency range of 500–600 kHz; energy parameter range is $0.0052 \leq E_{WT} \leq 0.0083$) and transversal (frequency range of 300–400 kHz; energy parameter range is $0.023 \leq E_{WT} \leq 0.094$ for ductile–brittle and $0.1 \leq E_{WT} \leq 0.15$ for brittle fracture) cracks in the fiber.

Comparing the parameters of AE signals recorded under tension of the composites of types 1 and 2, we can see that in the first case the values of the energy parameter of AE signals are higher ($0.00219 \leq E_{WT} \leq 0.708$) and are evenly distributed throughout the range (Fig. 4.14a) compared to the second case ($0.00028 \leq E_{WT} \leq 0.691$), where the predominant number of AE signals according to the criteria parameter is concentrated in the range of $0.00028 \leq E_{WT} \leq 0.24$ (Fig. 4.17a). Thus, the aramid fiber treatment with a silicone oil increases the adhesion between the fiber and the composite matrix, which is manifested by the change in the fracture character and decrease in the energy parameter of AE signals.

4.2.8 Characteristics of AE Signals During Tension of the Composite of Type 3

Figure 4.18 illustrates a typical AE signal during fracture of this composite. The signals have different amplitudes, but, as the spectral analysis shows, their dominant frequencies are in the range of 300–350 kHz. During the CM fracture, they alternate with signals in which, according to their spectral analysis, the dominant frequencies are in the range of 200–250 kHz.

According to the energy distribution at the decomposition levels, all the registered signals during fracture of the type 3 composite are divided into two groups (Fig. 4.19). The first one has the maximum energy at the level of D3 (250–500 kHz)—Type I. The second group has the maximum energy at level of D4 (125–250 kHz)—Type II. This coincides with their previously performed spectral analysis. The number of AE signals of Type II is 30% of the total number of the registered ones. Thus, after fiber treatment in plasma and polyethyleneimine with the addition of 10% nanoparticles, the ranges of the dominant frequency change compared to the composite of type 1, where the maximum energy is concentrated at the levels of D2 and D3.

Figure 4.20 shows the time distribution of AE signals registered during the experiment by the energy parameter, dominant frequency, and duration of radiation.

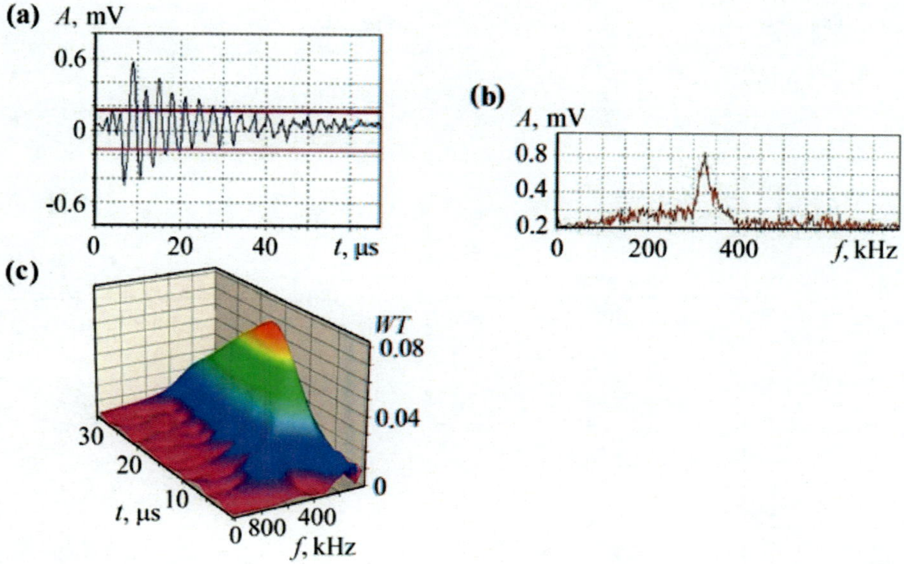

Fig. 4.18 Waveform (**a**), spectral characteristics (**b**), and CWT (**c**) of a typical AE signal during fracture of the composite of type 3 reinforced with an aramid fiber

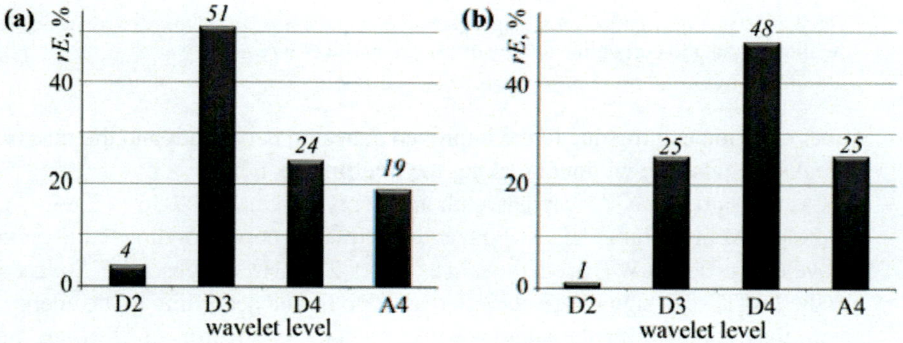

Fig. 4.19 Diagrams of the energy distribution by the levels of DWT decomposition of two types of AE signals during tension of the composite of type 3: **a** Type I; **b** Type II

The initial stages of fracture are characterized by AE signals with energy parameter $0.01 \leq E_{WT} \leq 0.042$, duration Δt up to 10 μs and dominant frequency in the range of 300–320 kHz. With further development of fracture, their energy parameter ($0.136 \leq E_{WT} \leq 0.178$) and duration Δt (up to 15 μs) increase. That is, according to the DWT, such signals are classified as Type I, and according to the energy criterion, they characterize the ductile–brittle and brittle fracture. According to the above identification, they accompany the propagation of cracks (longitudinal and transverse) in the aramid fiber. As found in [55], during tension of a composite with an aramid

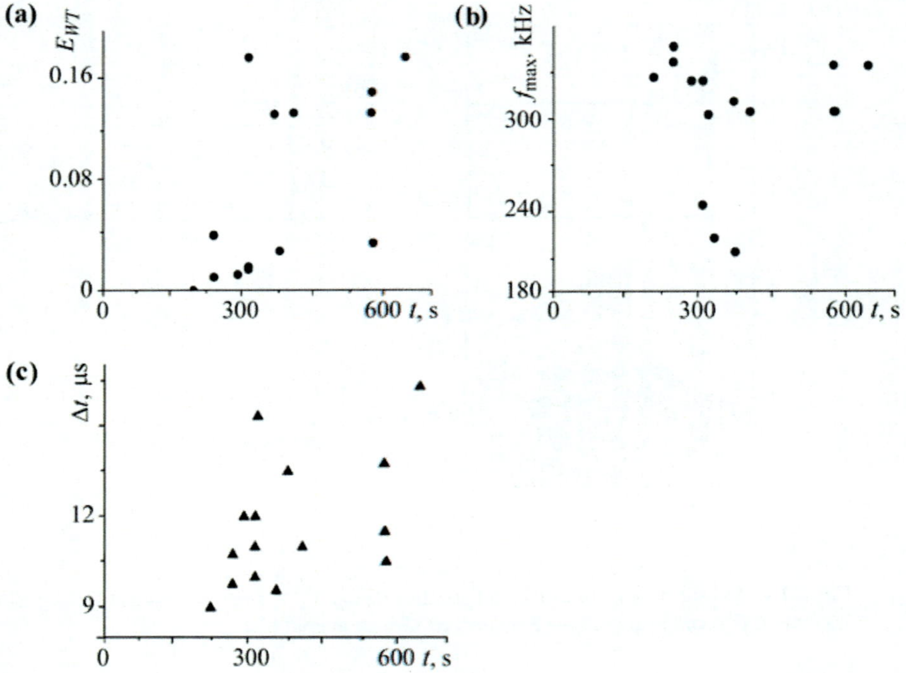

Fig. 4.20 Time distribution of AE signals by energy parameter (**a**), dominant frequency (**b**), and duration of radiation (**c**) during tension of the composite of type 3

fiber after ion etching due to the improved adhesion between it and the matrix, fiber breaks are possible without breaking the fiber/matrix bond.

At the same time, AE signals with an energy parameter $0.036 \leq E_{WT} \leq 0.048$, duration Δt in the range of 9–12 μs, and a frequency corresponding to the maximum wavelet coefficient WT_{max}, in the range of 200–250 kHz, are observed. So, according to the DWT, such signals are classified as Type II, and according to the energy criterion, they characterize the ductile–brittle fracture. In accordance with our method, they accompany the delamination of the fiber from the matrix. Since after fiber treatment in plasma and polyethyleneimine with the addition of 10% nanoparticles, the fracture toughness of the composite of type 3 increases compared to the composite of type 1, the energy parameter of the AE signals decreases.

At the end of the samples fracture, AE signals that have more than one spectral maximum in a short time interval are recorded. In Fig. 4.21, the CWT of one of them is presented. According to the parameters of CWT and DWT the event with the value of WT_{max1} is similar to the AE signals of type I, though with a higher energy parameter $E_{WT} = 0.195$. It is probably generated as a result of the rapid crack propagation through the cross section of the sample, which leads to its division into two parts. Taking into account the literature data [38, 56, 57], we can assume that event with WT_{max3} corresponds to the propagation of cracks in the cross section of the matrix, since the frequency at which this maximum is reached is 60 kHz.

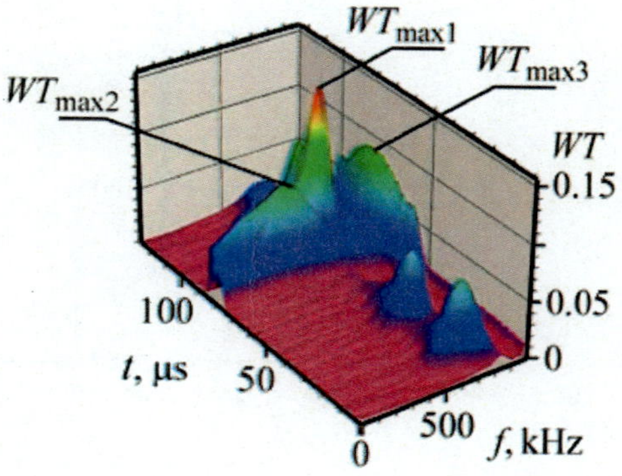

Fig. 4.21 CWT of AE signals at the final stage of fracture of the composite of type 3

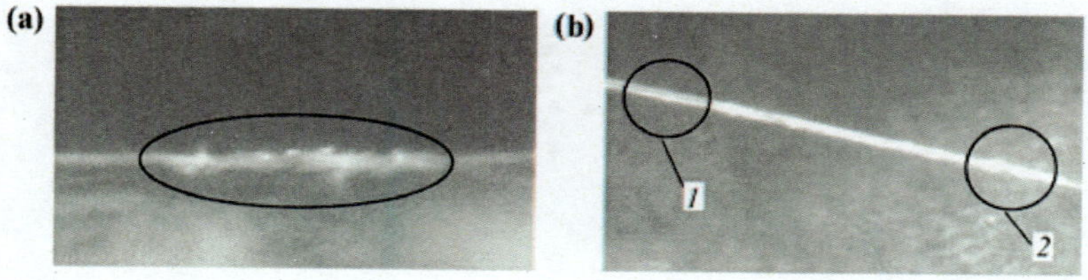

Fig. 4.22 Fracture of the composite of type 3 under the action of quasi-static tension: **a** fiber fracture; **b** delamination of the fiber from the matrix (section 1) and its rupture (section 2)

WT_{max2} is characterized by a wide frequency band, and the energy parameter (E_{WT} = 0.018) corresponds to the ductile–brittle fracture. In our case, such a signal can be generated, in particular, during the development of diagonal shears of the matrix parts [48] before dividing the sample into two pieces.

Figure 4.22 presents the appearance of the elements of the fractured sample of the composite of type 3, which confirm the correctness of the method of identifying the fracture mechanisms in such material. Figure 4.22a outlines the area of the destroyed fiber, which has not yet broken, and Fig. 4.22b shows its rupture in the zone of delamination from the matrix (section 2) and the propagation of delamination along the fiber without rupture (section 1).

4.2.9 Parameters of AE Signals During Tension of the Composite of Type 4

Figure 4.23 presents the time distribution of AE signals registered during the experiment according to the energy parameter, dominant frequency, and duration of radiation. Contrary to the previous case, during tension of this composite, one band of frequencies 290–320 kHz prevails. This indicates that one mechanism of fracture

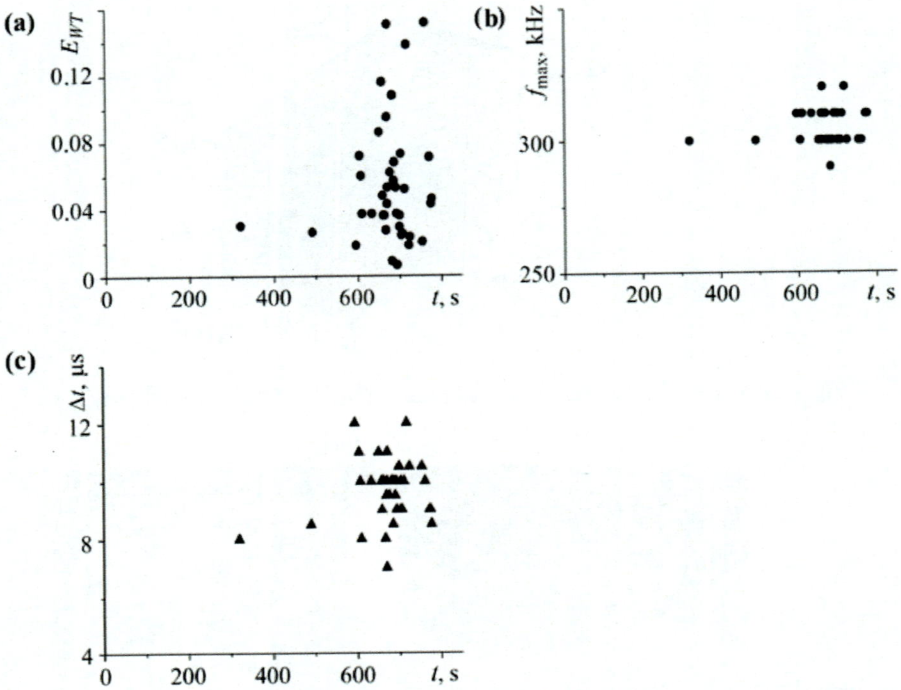

Fig. 4.23 Time distribution of AE signals by energy parameter (**a**), dominant frequency (**b**), and duration of radiation (**c**) during tension of the composite of type 4

predominates. In accordance with our identification method, AE signals accompany the propagation of longitudinal and transverse cracks in the fiber. Because of the specific feature of the aramid fiber treatment in plasma and polyethyleneimine with the addition of 100% nanoparticles, the adhesion between the fiber and the matrix is significantly improved, new bonds are formed between the fiber, nanoparticles, and matrix. This leads to a change in the character of the composite fracture.

With increasing load in the second half of the experiment, the AE activity increases significantly. At this time, AE signals with a dominant frequency band of 290–320 kHz and duration Δt in the range of 6–12 µs predominate, and the values of the energy parameter are in the range of $0.0064 \leq E_{WT} \leq 0.151$ that includes the range of ductile, ductile–brittle, and brittle types of fracture.

According to the developed method, the parameters of local events are set (maximum wavelet coefficient WT_{max}; frequency f_{max} corresponding to it; time t_{max} of occurrence of event, pulse frequency bandwidth Δf, duration of radiation Δt, energy parameter of local pulse E_{WT}), which are given in Table 4.3.

At the end of the experiment, AE signals are recorded with several local maxima, which are generated in a short time interval and accompany various mechanisms of composite fracture. Figure 4.24 shows the CWT of such a signal.

Analyzing the parameters of local events and taking into account the data of literature sources, it can be assumed that the local maximum WT_{max1} characterizes

Table 4.3 Amplitude–frequency and energy parameters of local maxima of AE signal for the composite of type 4 (see Fig. 4.23)

AE event	WT_{max}	f_{max}, kHZ	t_{max}, μs	Δf, kHZ	Δt, μs	E_{WT}
WT_{max1}	0.09	590	12.13	380	4	0.024
WT_{max2}	0.098	120	12.63	60	22	0.135
WT_{max3}	0.148	310	17.63	110	15	0.203

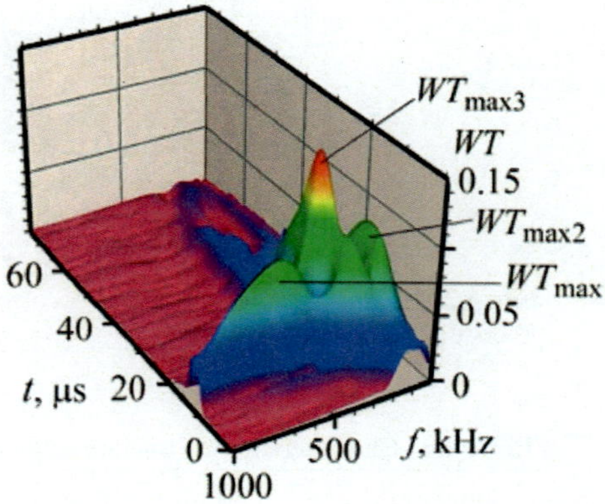

Fig. 4.24 CWT of AE signals at the final stage of fracture of the composite of type 4

the ductile–brittle fracture. The last one possibly has been generated during the development of diagonal shears of the matrix. The events with WT_{max2} and WT_{max3} correspond to crack propagation through the cross section of the matrix up to the division of the sample into two parts.

Figure 4.25 illustrates the distribution of local events, characterizing different types of fracture in composites. Regardless of the type of fiber treatment, the number of AE signals corresponding to brittle fracture decreases, while to ductile–brittle—increases compared to the tension of the composite with an aramid fiber without treatment. In particular, after treatment in a silicone oil, the percent of brittle fracture decreases by 60%, in plasma with a 10% coated fiber surface with nanoparticles—by 20%, in plasma with a 100% coated fiber surface with nanoparticles—by 80%. At the same time, the percent of ductile–brittle fracture increases in the first case by 30%, in the second—by 20%, in the third—twice. A composite in which the aramid fiber treated in the medium of plasma and polyethyleneimine with a 100% coated fiber surface with nanoparticles has the smallest percent of brittle and the largest percent of ductile–brittle fracture, thus indicating its greatest fracture resistance.

Therefore, using the analysis of the parameters of CWT and DWT of AE signals under tension of the aramid fiber-reinforced composite, a correspondence is established between the mechanisms and types of fracture, parameters of AE signals that accompany them (see Table 4.4).

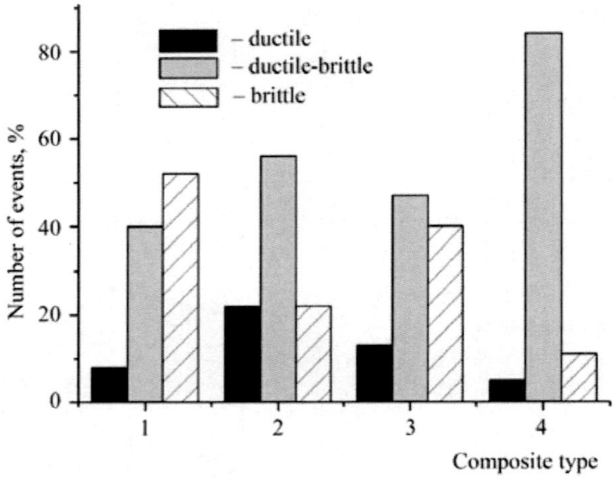

Fig. 4.25 Distribution of local pulses corresponding to different types of fracture of the aramid fiber-reinforced composites

Table 4.4 Characteristics of fracture of the aramid fiber-reinforced composites

Dominant fracture mechanism	Fracture type	Dominant frequency range, kHz	Duration of radiation, μs
Fiber/matrix delamination	Ductile–brittle, brittle	200–300	10–25
Propagation of longitudinal and transverse cracks in the fiber	Ductile, ductile–brittle, brittle	300–550	4–15
Propagation of cracks in the matrix	Brittle	<100	25–40
Shear mechanisms in the matrix	Ductile	500–600	5–10

References

1. Cooper, G. A. (1971). The structure and mechanical properties of composite materials. *Review of Physics in Technology., 2*(2), 49–91.
2. Gibson, R. F. (2010). A review of recent research on mechanics of multifunctional composite materials and structures. *Composite Structures, 92*(12), 2793–2810.
3. Dandekar, C. R., & Shin, Y. C. (2012). Modeling of machining of composite mateials: A review. *International Journal of Machine Tools and Manufacture, 57*, 102–121.
4. Subramani, N., Murali, J. G., Suresh, P., & Sankar, V. V. A. (2017). Review on hybrid composite materials and its applications. *International Research Journal of Engineering and Technology, 4*(2), 1921–1925.
5. Sun, X., Sun, H., Li, H., & Peng, H. (2013). Developing polymer composite materials: Carbon nanotubes or graphene? *Advanced Materials, 25*(37), 5153–5176.

6. Mohammed, L., Ansari, M. N. M., Pua, G., Jawaid, M., & Islam M. S. (2015). A review on natural fiber reinforced polymer composite and its applications. *International Journal of Polymer Science*. ID 243947. https://www.hindawi.com/journals/ijps/2015/243927
7. Gholizadeh, S. (2016). A review of non-destructive testing methods of composite materials. *Procedia Structural Integrity, 1*, 50–57.
8. Bozhydarnik, V. V., Skalskyi, V. R., & Matviiv, Y. Y. (2012). *Diahnostuvannya ruinuvannya sklovolokonnych kompozytiv metodom akustychnoyi emisiyi (Diagnistics of fiber reinforced composites fracture by acoustic emission method)*. Naukova dumka, Kyiv
9. De Rosa, I. M., Santulli, C., & Sarasini, F. (2009). Acoustic emission for monitoring the mechanical behaviour of natural fibre composites: A literature review. *Composites Part A Applied Science and Manufacturing, 40*, 1456–1469.
10. Dahmene, F., Yaacoubi, S., & Mountassir, M. E. (2015). Acoustic emission of composites structures: Story, success, and challenges. *Physics Procedia, 70*, 599–603.
11. Friedrich, K. (Ed.). (1989). *Application of fracture mechanics to composite materials. Composite materials series* (Vol. 6). Elsevier Science Publishers.
12. Guagliano, M., & Aliabadi, M. H. (Eds.). (2005). *Fracture and damage of composites*. WIT Press.
13. Talreja, R., & Singh, C. V. (2012). *Damage and failure of composite materials*. Cambridge University Press.
14. Cherepanov, G. P. (1983). *Mechanika razruscheniya kompozicyonnych materialov (Fracture mechanics of composite materials)*. Nauka.
15. Skal'skii, V. R., Stankevich, E. M., & Matviiv, Y. Y. (2013). A study of the features of the macrofracturing of composite materials. *Russian Journal of Nondestructive Testing, 49*(10), 562–571.
16. Skalsky, V., Stankevych, O., & Kuz, I. S. (2018). Application of wavelet transforms for the analysis of acoustic-emission signals accompanying fracture processes in materials (A survey). *Material Science, 54*(2), 139–153.
17. Jonson, M. (2003). Classification of AE transients based on numerical simulations of composite laminates. *NDT E International, 36*, 319–329.
18. Qi, G., Barhorst, A., Hashemi, J., & Kamala, G. (1997). Discrete wavelet decomposition of acoustic emission signals from carbon-fiber-reinforced composite. *Composites Science and Technology, 57*, 389–403.
19. Kalogiannakis, G., Quintelier, J., De Baets, P., Degrieck, J., & Van Hemelrijck, D. (2008). Identification of wear mechanisms of glass/polyester composites by means of acoustic emission. *Wear, 264*, 235–244.
20. Marec, A., Thomas, J.-H., & El Guerjouma, R. (2008). Damage characterization of polymer-based composite materials: Multivariable analysis and wavelet transform for clustering acoustic emission data. *Mechanical System and Signal Processing, 22*, 1441–1464.
21. Loutas, T. H., Kostopoulos, V., Ramirez-Jimenez, C., & Pharaoh, M. (2006). Damage evolution in center-holed glass/polyester composites under quasi-static loading using time/frequency analysis of acoustic emission monitored waveforms. *Composites Science and Technology, 66*, 1366–1375.
22. Velayudham, A., Krishnamurthy, R., & Soundarapandian, T. (2005). Acoustic emission based drill condition monitoring during drilling of glass/phenolic polymeric composite using wavelet packet transform. *Materials Science and Engineering A, 412*, 141–145.
23. Wang, X., Xiang, J., Hu, H., Xie, W., & Li, X. (2015). Acoustic emission detection for mass fractions of materials based on wavelet packet technology. *Ultrasonic, 60*, 27–32.
24. Grabowska, J., Palacz, M., & Krawczuk, M. (2008). Damage identification by wavelet analysis. *Mechanical System and Signal Processing, 22*, 1623–1635.
25. Coifman, R. R., & Wickerhauser, M. V. (1992). Entropy-based algorithms for best basis selection. *IEEE Transactions on Information Theory, 38*(2), 713–718.
26. Rafiee, J., Tse, P. W., Harifi, A., & Sadeghi, M. H. (2009). A novel technique for selecting mother wavelet function using an intelligent fault diagnosis system. *Expert Systems with Application, 36*, 4862–4875.

27. Wang, X. H., Zhu, C. M., Mao, H. L., & Huang, Z. F. (2009). Wavelet packet analysis for the propagation of acoustic emission signals across turbine runners. *NDT E International, 42*, 42–46.
28. Li, Y., & Yi-Chu, Z. (2006). Wavelet analysis of acoustic emission signals from thermal barrier coatings. *Transactions of Nonferrous Metals of Society of China, 16*, 270–275.
29. Stepanova, L. N., Ramazanov, I. S., Kabanov, S. I. et al. (2007). Ispolzovaniye veivlet-filtracii pri loaklozacii signalov akusticheskoi emissii (Use of wavelet filtering for localization of acoustic emission signals). *Kontrol. Diagnostika (Contrlol. Diagnostics), 9*, 27–31.
30. Khamedi, R., Fallahi, A., & Oskouei, A. R. (2010). Effect of martensite phase volume fraction on acoustic emission signals using wavelet packet analysis during tensile loading of dual phase steels. *Materials & Design, 31*, 2752–2759.
31. Sung, D.-U., Kim, C.-G., & Hong, C.-S. (2002). Monitoring of impact damages in laminates using wavelet transform. *Composites Part B Engineering, 33*, 35–43.
32. Velayudham, A., Krishnamurthy, R., & Soundarapandian, T. (2005). Acoustic emission based drill condition monitoring during drilling of glass/phenolic polymeric composite using wavelet packet transform. *Materials Science and Engineering A, 412*, 141–145.
33. Tang, S., Tong, M., Hu, J., & He, X. (2010). Characteristics of acoustic emission signals in damp cracking coal rocks. *Mining Science and Technology, 20*, 143–147.
34. Qi, G. (2000). Wavelet-based AE characterization of composite materials. *NDT E International, 33*, 133–144.
35. *SKOP-8M portable 8-channel acoustic emission device for non-destructive testing of materials, products and structure elements*. Retrieved 25 March, 2021, from https://www.ipm.lviv.ua/our_developments/our_developments.php
36. Ni, Q.-Q., & Iwamoto, M. (2002). Wavelet transform of acoustic emission signals in failure of model composites. *Engineering Fracture Mechanics, 69*, 717–728.
37. Gutkin, R., Green, C. J., Vangrattanachai, S., Pinho, S. T., Robinson, P., & Curtis, P. T. (2011). On acoustic emission for failure investigation in CFRP: Pattern recognition and peak frequency analyses. *Mechanical System and Signal Processing, 25*, 1393–1407.
38. Singh, T. J., & Samanta, S. (2015). Characterization of Kevlar fiber and its composites: A Review. *Materials Today: Proceedings, 2*(4/5), 1381–1387.
39. Dai, S.-R., & Piggott, M. R. (1993). The strengths of carbon and Kevlar fibres as a funcion of their lengths. *Composites Science and Technology, 49*, 81–87.
40. Morgan, R. G., Pruneda, C. O., & Steele, W. J. (1983). The relationship between the physical structure and the microscopic deformation and failure processes of poly (p-phenylene terephthalamide) Fibers (Kevlar-49). *Journal of Polymer Science, 21*(9), 1757–1783.
41. Hill, R., & Okoroafor, E. U. (1994). Surface conditions of Kevlar and glass and mechanical properties studied by Weibull statistics and acoustic emission. *Composites, 25*(10), 913–916.
42. Yasuda, H. (1976). Plasma for modification of polymers. *Journal of Macromolecular Science Part A Chemistry, 3*, 383–420.
43. Sheu, G. S., & Shyu, S. S. (1994). Surface properties and interfacial adhesion studies of aramid fibres modified by gas plasmas. *Composites Science and Technology, 52*(4), 489–497.
44. Ren, Y., Wang, C., & Qiu, Y. (2007). Influence of aramid fiber moisture regain during atmospheric plasma treatment on aging of treatment effects on surface wettability and bonding strength to epoxy. *Applied Surface Science, 253*, 9283–9289.
45. Park, J.-M., Kim, D.-S., & Kim, S.-R. (2003). Improvement of interfacial adhesion and nondestructive damage evaluation for plasma-treated PBO and Kevlar fibers/epoxy composites using micromechanical techniques and surface wettability AE. *Journal of Colloid and Interface Science, 264*(2), 431–445.
46. Tittman, B. R., Miysaka, C., Guers, M., Kasano, H., & Morita, H. (2013). Non-destructive evaluation (NDE) of aerospace composites: acoustic microscopy. In V. M. Karbhari (Ed.), *Non-destructive evaluation (NDE) of Polymer Matrix Composites* (pp 423–448, 449e). Woodhead Publishing
47. Bunsell, A. R. (1975). The tensile and fatigue behavior of Kevlar 40 (PRD-49) fibre. *Journal of Materials Science, 10*(8), 1300–1308.

48. Davidovitz, M., Mittelman, A., Roman, I., & Marom, G. (1984). Failure modes and fracture mechanisms in flexure of Kevlar-epoxy compoites. *Journal of Materials Science, 19*(2), 377–384.
49. Rajan, D. S., Kishore, N. N., & Agarwal, B. D. (1989). Damage characterization in Kevlar/epoxy composites using acoustic emission technique. *Journal of Acoustic Emission, 8*(1/2), 5297–5299.
50. Caneva, C., De Rosa, I. M., & Sarasini, F. (2007). Acoustic emission monitoring of flexurally loaded aramid/epoxy composites by embedded PVDF sensors. *Journal of Acoustic Emission, 25*, 80–91.
51. Caneva, C., De Rosa, I. M., & Sarasini, F. (2008). Monitoring of impacted aramid-reinforced composites by embedded PVDF acoustic emission sensors. *Strain, 44*(4), 308–316.
52. Hamstad, M. A., & Moore, R. L. (1986). Acoustic emission from single and multiple Kevlar 49 filament breaks. *Journal of Composite Materials, 20*(1), 46–66.
53. Ma, B. T., Schadler, L. S., Laird, C., & Figueroa, J. C. (1990). Acoustic emission in single filament carbon/polycarbonate and Kevlar/poycarbonate composites under tensile deformation. *Polymer Composites, 11*(4), 211–216.
54. Mittelman, A., & Roman, I. (1991). Monitoring the tensile deformation in real unidirectional Kevlar-epoxy composites. *NDT E International, 24*(2), 85–89.
55. Narisawa, I., & Oba, T. (1985). An evaluation of acoustic emission from fibre-reinorced composites. Part 2. The application of acoustic emission techniques to aramid fibre-reiforced model composites. *Journal of Materials Science, 20*(12), 4527–4531.
56. Woo, S.-C., & Kim, T.-W. (2014). High-strain-rate impact in Kevlar-woven compoites and fracture analysis using acoustic emission. *Composites Part B Engineering, 50*, 125–136.
57. Woo, S.-C., & Kim, T.-W. (2016). High strain-rate failure in carbon/Kevlar hybrid woven composites via a novel SHPB-AE coupled test. *Composites Part B Engineering, 97*, 317–328.
58. Netravali, A. N., Li, Z.-F., Sachse, W., & Wu, H. F. (1991). Determination of fibre/matrix interfacial shear strength by an acoustic emission technique. *Journal of Materials Science, 26*(24), 6631–6638.

Chapter 5
Ranking of Dental Materials and Orthopedic Constructions by Their Tendency to Fracture

In modern orthopedic dentistry, three types of restorative materials are used: ceramics, metals, polymers [1]. Metals which possess high strength and rigidity are used to ensure the ability of denture to bear significant mechanical loads. At the same time, because of the denture ability to conduct heat quickly and the unaesthetic view, the use of such constructions is limited. Instead, ceramics and polymers, which are thermal insulators, having a light color and are translucent, have become widespread, which leads to their use for the protection of teeth from changes in oral temperature and for the creation of aesthetic fillings and dentures that restore the natural view of a human teeth. Combinations of different chemical materials are often practiced, as none of them alone can be considered the perfect.

When choosing the appropriate material for orthopedic restorations, it is necessary to have complete information about its strength. The last one can change under the influence of various factors (defects of material in critical areas; state of the restoration surface; increased loads on the construction, etc.).

For effective clinical use of certain materials, it is necessary not only to consider their mechanical properties, but also to know the dynamics of fracture processes. Such information can be obtained by using in mechanical tests the AE method with signal processing by WT.

This chapter considers the mechanical characteristics and peculiarities of fracture of dental polymers, ceramics, and composites; the express technique of their ranking by the energy criterion of the identification of the fracture types is constructed, and the peculiarities of fracture of the tooth-endocrown system under quasi-static compression are investigated.

V. Skalskyi et al., *Acoustic Emission*, Foundations of Engineering Mechanics,
https://doi.org/10.1007/978-3-031-11291-1_5

5.1 State of the Art of Researches on Mechanical Properties of Dental Materials

Such materials as biopolymers Ralavit-55, Revotek, RegularSet, Temphase, Protemp, Luxatemp, Trim, Akrodent, Structur, Tempron, etc., are most often used for temporary crowns and bridges [1, 2]. The mechanical strength of such materials affects the integrity of the constructions in application. Many studies of the mechanical properties of materials for temporary orthopedic constructions and their comparison are known in the literature. In particular, the bending strength is determined by simulating a different clinical situation from the beginning of manufacture of a temporary structure to a certain time of its use [2–12]; fatigue strength is evaluated [13–15]; the influence of various factors on the surface hardness of materials is established [16–19]; the efficiency of using polymers reinforced with different types of fibers in temporary constructions is studied [20–25]; the strength of temporary constructions after restoration [24] and their crack growth resistance are investigated [26, 27]; the effect of water absorption on the strength characteristics of dental composites is determined [28–30].

A literature review [31] summarizes the numerous results of studies of dental composites with different bases for temporary orthopedic constructions. All of them are damaged by the short- or long-term acting stresses and water sorption. So they need to be quickly replaced by permanent constructions or materials should be improved to ensure their best performance.

Since 1999, endocrowns have become widespread in the practice of orthopedic treatment [32]. They are made of various materials, primarily endocrowns of feldspar and lithium-silicate-reinforced ceramics, hybrid composites, the latest CAD/CAM ceramics, and composite blocks. Mechanical properties and the propagation of fracture in endocrown restorations are determined during mechanical tests [33–39] or by using the finite element analysis [40–45].

In the study of the properties of different types of dental materials and constructions, the AE method is successfully used. It provides real-time information about the development of damage or fracture. The peculiarities of composite shrinkage during polymerization [46–50] and fracture [51–56]; adhesion between metal and other ceramic materials in orthopedic restorations [57, 58]; destruction of polymers [59, 60] are studied.

Thus, analysis of literature sources shows that the study of the properties of dental materials for temporary and permanent fixed constructions as well as of the effectiveness of their operation is topical in modern dental materials science.

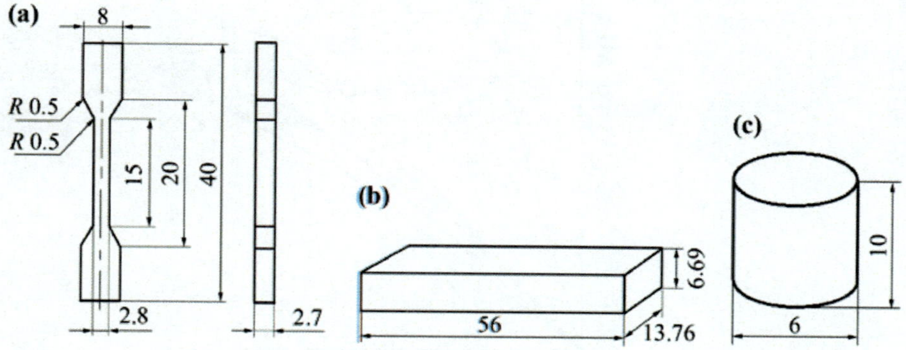

Fig. 5.1 Geometry of the lamellar (**a**), prismatic (**b**), and cylindrical (**c**) samples of dental polymers (mm)

5.2 Determination of the Characteristics of Materials for Temporary Fixed Constructions of Dentures

The initiation and propagation of fracture are studied and the characteristics of strength and plasticity, fracture toughness under quasi-static tension and compression, as well as the number of physical characteristics of polymeric provisional materials are evaluated here.

5.2.1 Materials for Investigations

Samples made of polymeric provisional materials Protemp 4 (3M ESPE, USA), Akrodent (JSC STOMA, Ukraine), Structur 2SC (VOCO, Germany), Tempron 1-1PKG (GC, Japan), Ceramill PMMA (AmannGirrbach, Austria) are tested. Three types of samples are prepared for experiments (Fig. 5.1).

The first type (plate) (Fig. 5.1a) is used to determine the tensile characteristics of strength and plasticity, static crack growth resistance, water absorption, and shrinkage; *the second type* (prismatic with the induced stress concentrators) (Fig. 5.1b)—to assess the static crack growth resistance under tension; *the third type (cylindrical)* (Fig. 5.1c)—to establish the strength characteristics of polymers under compression.

5.2.2 Evaluation of Strength Characteristics of Polymers

The samples are subjected to tension on a testing machine SVR-5 with a loading speed of 4×10^{-7} m/s.

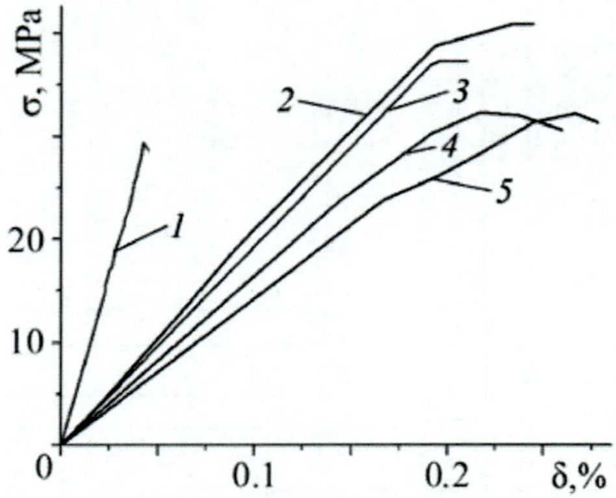

Fig. 5.2 Stress–strain curves of the samples of dental polymers: *1*—Protemp; *2*—Acrodent; *3*—Structur; *4*—Tempron; *5*—Ceramill

Table 5.1 Mechanical characteristics of dental polymers under tension

Polymer	$\sigma_{0.2}$, MPa	Place in the ranking	σ_B, MPa	Place in the ranking	δ, %	Place in the ranking
Protemp	38.9	1	40.9	1	0.244	3
Akrodent	37.3	2	37.3	2	0.209	4
Structur	30.4	3	32.3	3	0.26	2
Tempron	27.4	4	31.9	4	0.28	1
Ceramill	23.5	5	28	5	0.04	5

Figure 5.2 shows the combined stress–strain curves of the samples of different polymers.

Table 5.1 presents the average values obtained from the curves of strength and plasticity (yield strength $\sigma_{0.2}$ and strength σ_B, as well as the relative deformation δ) and ranked materials (from the best one) for each parameter.

Five samples of each polymer are prepared for *compression tests*, with employment of specially prepared indenters attached to the UME-10TM loading machine with a traverse speed of 0.1 mm/min. Typical compression curves of polymers are presented in Figs. 5.3, 5.4, and 5.5.

As we can see, the elastic characteristics of strength under tensile and compression loading differ significantly. Polymers have a significant elastic compression deformation, which many times exceeds their tensile elastic limit.

Therefore, compression stress σ_1, at which a curve diverges from linearity and corresponding to it relative deformation δ_1, and also maximum compression stresses σ_{max}, of the polymer as well as corresponding deformation δ_{max} are determined. Test results and the corresponding ranking for each material are presented in Table 5.2. Experimental data are averaged for each material.

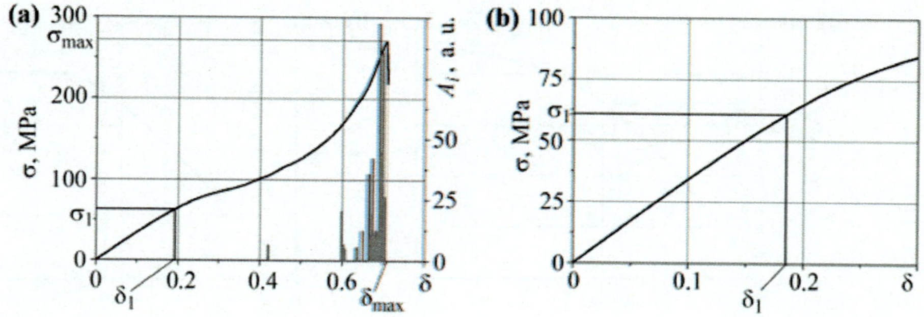

Fig. 5.3 Compression curve of a cylindrical sample of Akrodent polymer with the distribution of AE signals amplitude (**a**) and its elastic region (**b**)

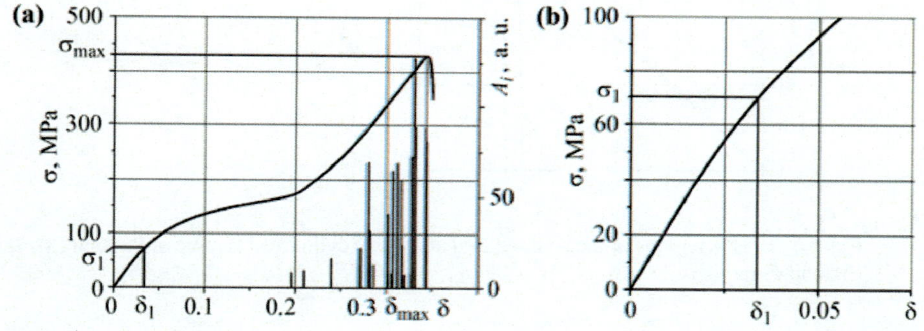

Fig. 5.4 Compression curve of a cylindrical sample of Protemp polymer with the distribution of AE signals amplitude (**a**) and its elastic region (**b**)

To assess the crack growth resistance of dental materials, the experiments are performed in two stages [60]: At first, we study the fracture of polymers during quasi-static tension of Type I samples. To confirm the reliability of the obtained values of fracture toughness, the Type II samples are tested under three-point bending according to the requirements of the State Standard GOST 25.506–85 [61].

Eight samples of each polymer are prepared for the first stage of the experiments (Fig. 5.6a). The samples prepared accordingly after polymerization are subjected to tension on the SVR-5 testing machine with a capture rate of 4×10^{-7} m/s (Fig. 5.6b).

Simultaneously with the tension of samples, the AE information is recorded using the detection system SKOP-8M [62]. Signals detected by the sensor were passed through the preamplifier of 40 dB gain with the band pass of 200–600 kHz. AE signals were recorded using 4 MHz sampling frequencies.

It is found (Fig. 5.7) that all the polymers are failed by the linear-elastic mechanism, except for Protemp, the fracture diagram of which has a plastic region. This result is consistent with the known studies [26, 27].

To evaluate the fracture types in polymers, CWT is constructed for AE signals obtained under initial loading (signal A), in the middle section of the fracture diagram

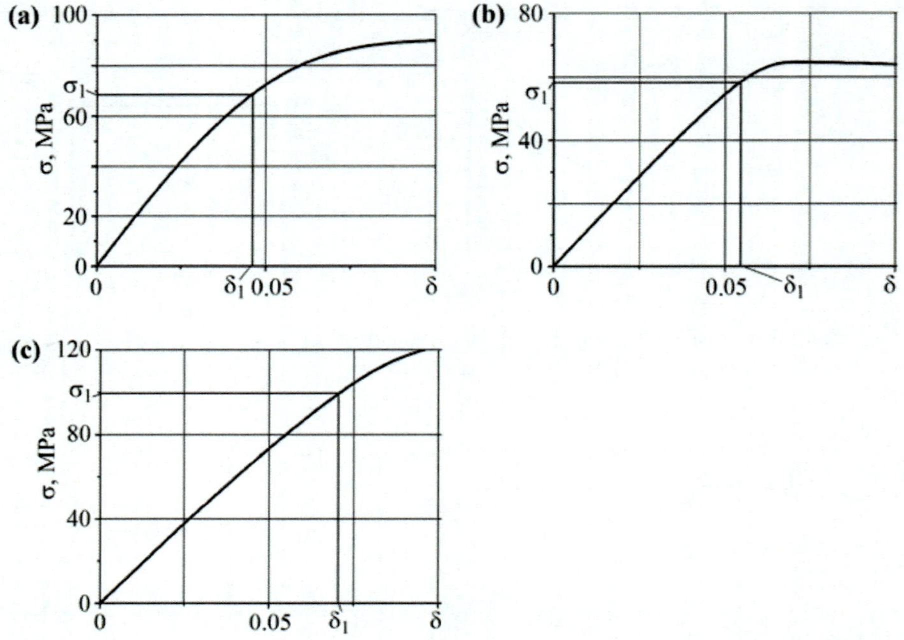

Fig. 5.5 Elastic regions of the compression curve of a cylindrical sample of Structur (**a**), Tempron (**b**), and Ceramill (**c**)

Table 5.2 Mechanical characteristics of polymers under compression

Polymer	σ_1, MPa	Place in the ranking	δ_1, %	Place in the ranking	σ_{max}, MPa	Place in the ranking	δ_{max}, %	Place in the ranking
Akrodent	73.5	2	19	1	305.41	3	71	1
Protemp	70.35	3	3.3	5	453.28	1	34	4
Structur	68.62	4	4.4	4	328.16	2	26	5
Tempron	58.09	5	6	3	121.38	5	69	2
Ceramill	99.1	1	7	2	131.62	4	44	3

(B) and when higher (critical) load values (C and D) are reached. The parameter E_{WT} is determined by relation (3.3).

The SIF K_{I} is given by the formula:

$$K_{\mathrm{I}} = \sigma\sqrt{\pi l} \cdot Y, \tag{5.1}$$

where $Y = \left[\frac{2b}{\pi l}\left(\mathrm{tg}\frac{\pi l}{2b} + 0.1\sin\frac{\pi l}{b}\right)\right]^{1/2}$ is the correction factor [63].

The results of calculations are presented in Table 5.3. According to the energy criterion for identifying the types of structural materials fracture in the initial stages,

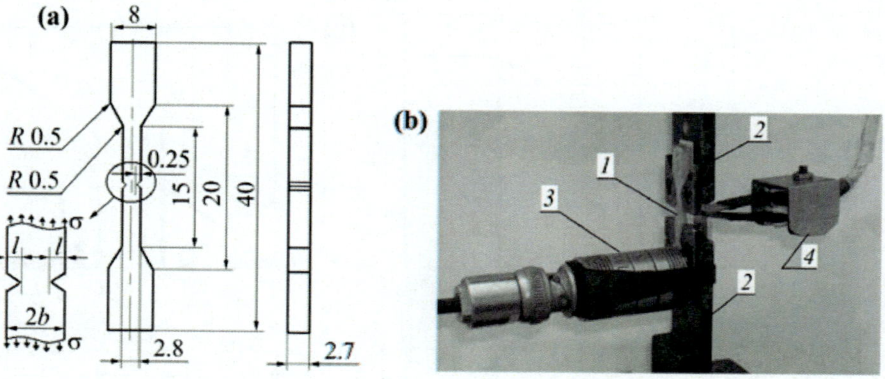

Fig. 5.6 Geometry of the sample for testing of polymer materials (mm) **a** and its general view in the grips of the installation during tension **b**: *1*—sample; *2*—grip; *3*—AE sensor; *4*—primary converter of the crack opening displacement

in most polymers (Protemp, Akrodent, Structur) AE signals are generated, which correspond to the ductile–brittle fracture ($0.01 \leq E_{WT} < 0.1$). In the Tempron and Ceramill materials, the brittle fracture mechanisms prevail throughout the loading time. This is evidenced by the high values of the criterion parameter, although plastic deformation is also present.

As one can see from Table 5.3, the presence of several local maxima of wavelet coefficients with different frequencies is a specific feature of CWT of AE signals that accompany the fracture of Akrodent, Tempron, Ceramill materials. This can mean the alternation of different fracture events, which by the values of the criteria parameter, correspond to the ductile, brittle, or ductile–brittle types of fracture.

Figure 5.8 shows the CWT of signal D from Fig. 5.7b, which has several local maxima corresponding to the ductile (local maximum I) and ductile–brittle (local maxima II and III) fracture of the provisional material Akrodent.

For the event D according to the CWT (Fig. 5.8) at time $t = 10.25$ μs the ductile fracture $E_{WT} = 0.0001$ (local maximum I) occurs. In this case, $WT_{max} = 0.0385$, $\Delta f = 0.3$ MHz, the frequency corresponding to WT_{max} of $f_{max} = 650$ kHz. Then, at time $t = 11.75$ μs, the ductile–brittle fracture starts (local maximum II), which hypothetically may correspond to the growth and coalescence of microcracks. Parameters of the local maximum II are as follows: $WT_{max} = 0.036$, $\Delta f = 0.105$ MHz, $f_{max} = 270$ kHz. Local maximum III, which reaches a peak at time $t = 24.25$ μs ($WT_{max} = 0.058$, $\Delta f = 0.125$ MHz, $f_{max} = 330$ kHz, $E_{WT} = 0.025$), characterizes the further development of microcracking with the transition to macrocrack formation, which eventually leads to the sample failure.

Note that the radiation of AE signals, which means the beginning of a defect formation in the material, occurs under different loading of each dental polymer sample. According to the experimental results, provisional materials by their ability to resist crack initiation can be ranked in the following order (from the strongest): Protemp,

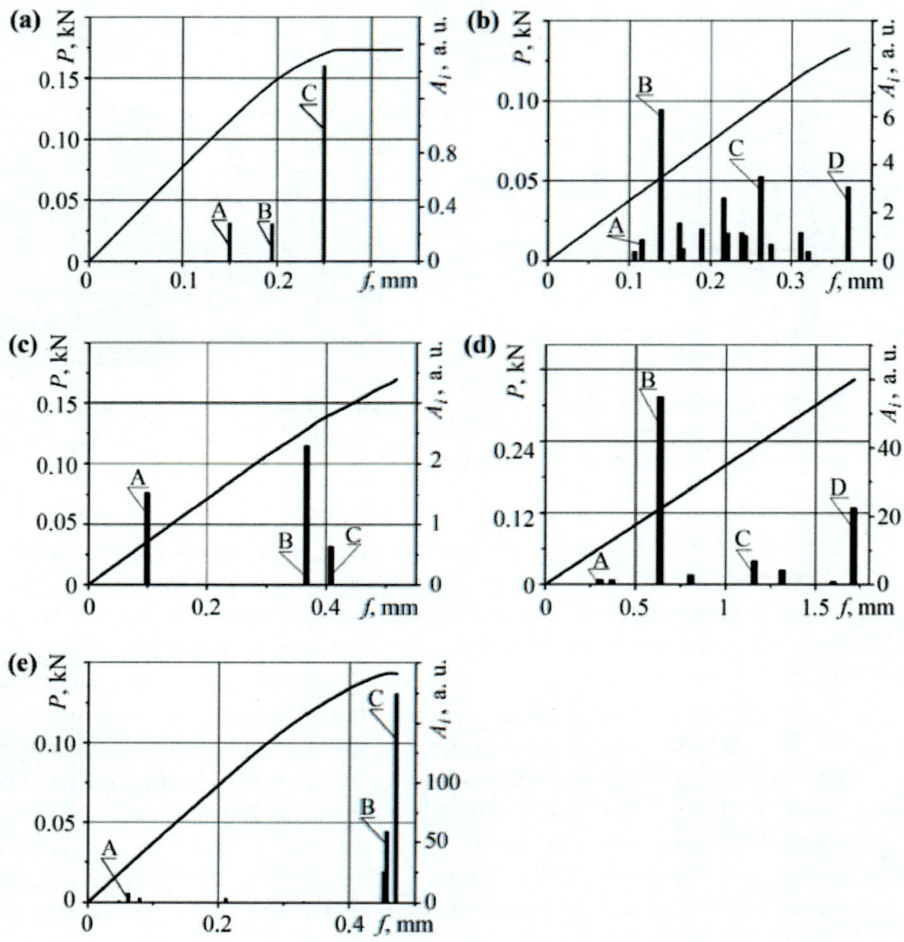

Fig. 5.7 Diagrams of fracture and distribution of AE signals amplitude for the samples of polymer materials Protemp (**a**), Akrodent (**b**), Structur (**c**), Ceramill (**d**), and Tempron (**e**): A, B, C, D are AE signals under different loads

Akrodent, Structur, Tempron, Ceramill, and by the macroindex of fracture toughness (from the most ductile)—Protemp, Tempron, Structur, Akrodent, Ceramill.

It can be stated by the analysis of fracture diagrams and the values of the criterion parameter of fracture types E_{WT}, that most of the studied dental polymers fracture brittle by the linear-elastic mechanism, and Protemp—by the elastic–plastic one. Protemp demonstrates the highest crack growth resistance and Ceramill—the lowest.

In the second stage of the tests, first the stress notches are made on the samples. Then fatigue cracks are induced under cyclic loading with a frequency of 17 Hz, so that at the final stage their growth rate is the lowest. Table 5.4 presents experimental data on induction of fatigue cracks from stress concentrators on the polymer samples, and Table 5.5 summarizes the results of fatigue crack growth resistance tests, obtained

Table 5.3 Averaged values of the criteria parameter E_{WT} and crack growth resistance characteristics for the corresponding AE signals

Polymer	A		B		C		D		K_{IC}, MPa·m$^{1/2}$
	E_{WT}	K_I, MPa·m$^{1/2}$	E_{WT}	K_I, MPa·m$^{1/2}$	E_{WT}	K_I, MPa·m$^{1/2}$	E_{WT}	K_I, MPa·m$^{1/2}$	
Protemp	0.01	0.95	0.005	1.05	0.06	1.4	–	–	1.4
Akrodent	0.015	0.27	0.146	0.44	0.015	0.65	0.001 0.01 0.025	0.91	0.91
Structur	0.017	0.2	0.07	0.74	0.007	0.97	–	–	0.97
Tempron	0.214	0.16	0.013 0.0014	0.99	0.449	–	–	–	0.99
Ceramill	0.08	0.15	0.13 0.326	0.31	0.0055	0.52	0.005 0.01	0.82	0.82

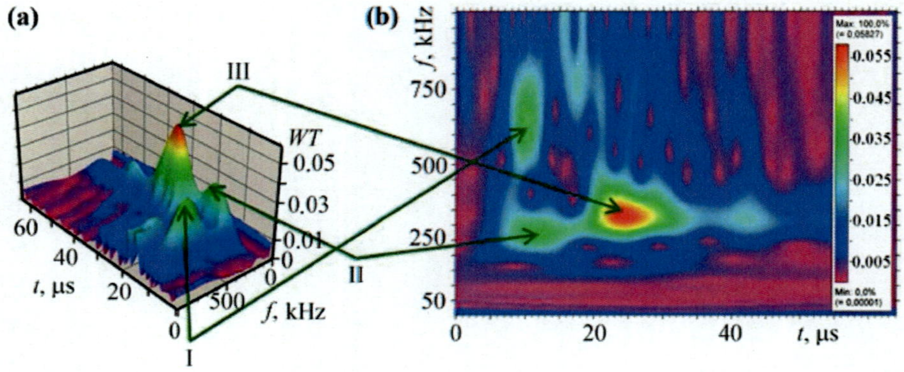

Fig. 5.8 CWT of AE signal (event D in Fig. 5.7b) during fracture of the Akrodent polymer sample (**a** 3D plot of wavelet spectrum intensity; **b** 2D frequency projection)

Table 5.4 Characteristics of cyclic loading during fatigue cracks induction

Polymer	Number of loading cycles *N*, thous	Crack propagation rate at the final stage *v*, mm/cycle	Maximum stresses in a cycle σ_{max}, MPa	Stress ratio *R*
Protemp	537	1.373×10^{-6}	0.058	0.06
Akrodent	1443	1.25×10^{-6}	0.054	0.003
Structur	154	3.12×10^{-6}	0.073	0.09
Tempron	142	1.01×10^{-6}	0.052	0.06
Ceramill	170	2.68×10^{-6}	0.062	0.09

by the Standard GOST 25.506–85 [61], and for each case the polymers have been ranked.

To calculate the SIF K_{IAE} of AE signals radiation, their maximum values $K_{IAE}{}^{max}$, as well as the critical value K_{IC} (the index of static crack growth resistance) we use the following formula [61]:

$$K_I = (PL/\mathrm{tb}^{3/2})Y_4;\ Y_4 = 3.494(1 - 3.396\lambda + 5.839\lambda^2);\ \lambda = l/b, \quad (5.2)$$

where P is the load; L is the distance between the supports; t is the sample thickness; b is the sample height; Y_4 is the correction function; l is the crack length.

The obtained data are checked by the Srawley–Gross formula:

$$K_I = (1.5PL/\mathrm{tb}^{3/2})Y_4;\ Y_4 = \lambda^{1/2}3.494(1 - 3.396\lambda + 5.839\lambda^2). \quad (5.3)$$

The calculation results coincide with an accuracy of 1%.

As one can see, the results obtained under different modes of loading and using different formulas for calculating static crack growth resistance are somewhat

Table 5.5 Characteristics of AE activity and fracture toughness of polymers under different loading modes and calculation methods

Polymer	$\Sigma A_i \times 10^{-3}$ a.u	Place in the ranking	Crack growth resistance characteristics, MPa·m$^{1/2}$					
			K_{IAE}	Place in the ranking	K_{IAE}^{max}	Place in the ranking	K_{IC}	Place in the ranking
Tension								
Akrodent	38.33	3	0.12	2	0.24	4	0.41	4
			0.27[a]		0.41[a]		0.91[a]	
Protemp	1.73	1	0.32	1	0.40	1	0.42	1
			0.95[a]		1.4[a]		1.42[a]	
Structur	4.45	2	0.11	3	0.39	3	0,51	3
			0.2[a]		0.74[a]		0.97[a]	
Tempron	236.59	4	0.07	4	0.44	2	0.44	2
			0.16[a]		1.01[a]		1.01[a]	
Ceramill	1390.5	5	0.09	5	0.23	5	0.58	5
			0.15[a]		0.31[a]		0.82[a]	
Three-point bending								
Protemp	23.25	1	0.268	3	1.34	1	1.134	3
Akrodent	114.77	4	0.29	2	1.05	2	1.72	1
Structur	44.85	2	0.23	4	0.83	3	0.83	5
Tempron	325.5	5	0.38	1	0.79	4	1.67	2
Ceramill	81.27	3	0.171	5	0.95	5	1.1	4

[a] data obtained by Irwin's formula

different for the same loads and crack lengths. This is due to a large-scale factor and different development of the plastic zone at the macrocrack tip. The values obtained by formulas (5.2) or (5.3) are the most reliable under three-point bending. Under tension of samples the data obtained by Irwin's formula (5.1) for the plates with two lateral cracks [63] are the most trustworthy. The obtained results agree well with known literature studies [26, 27, 64–66].

5.2.3 *Evaluation of Water Absorption of Polymer Materials*

For this purpose, Type II samples (10 pieces each) are prepared from polymer materials according to the manufacturer's instructions. Then they are aged in air for 6648 h. Prior to testing, the samples are kept in a thermostat TC-80-U42 for 72 h at a temperature of 37.0 ± 1.0 °C, then transferred to a desiccator for 60 min, later weighed (m_0) with an accuracy of 0.0001 g on electronic scale RADWAG WAA-160. At last, the samples are exposed to a physiological NaCl solution (9 g NaCl per 1 l of water) in a

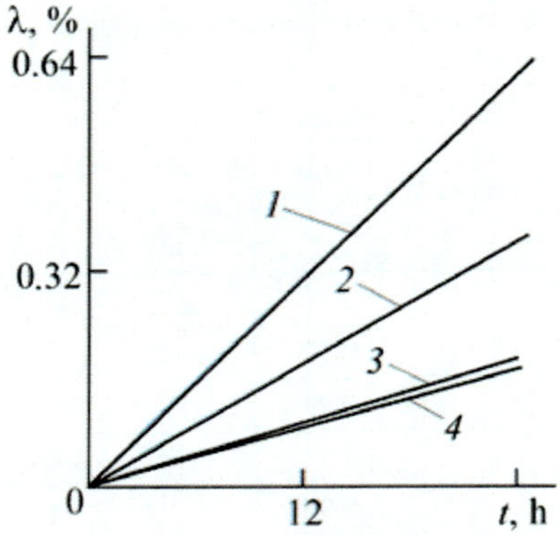

Fig. 5.9 Time dependence of water absorption of polymers: *1*—Akrodent; *2*—Tempron; *3*—Protemp; *4*—Structur

thermostat at a temperature of 37 ± 1 °C for twenty-four hours, since this is the time for the most intensive water absorption. After exposure, the samples are removed, dried with lint-free filter paper, and reweighed.

The degree of water absorption (λ, %) is calculated by the formula

$$\lambda = 100(m_1 - m_0)/m_0, \tag{5.4}$$

where m_0 is the weight of the sample before immersion in a NaCl solution (g); m_1 is the sample weight after being in the solution (g).

The results of evaluating the degree of water absorption of materials are processed by the statistical analysis methods for $p = 90\%$ in the program STATISTICA v.6.0 (Fig. 5.9).

Absorption occurs within 24 h; i.e., the test samples are saturated with liquid when the physiological solution fills the micropores of the polymer samples surface. The absorption of each material is different.

The Structur and Protemp samples have the lowest degree of water absorption after 24 h exposure to the physiological solution (from 0.180 ± 0.0032 to 0.192 ± 0.0038% and from 0.193 ± 0.0032 to 0.201 ± 0.0038%, respectively). The Tempron and Ceramill samples (approximately from 0.368 ± 0.0511 to 0.378 ± 0.0298%) have the intermediate degree, and the Akrodent samples—the highest one (from 0.638 ± 0.412 to 0.649 ± 0.0225%).

Table 5.6 presents the degrees of water absorption of materials, calculated by relation (5.4), as well as their ranking starting from the lowest.

Note that the conditions of laboratory tests differ from the real use of dentures in oral cavity, where the elastic materials are subjected to the influence of cyclic masticatory loads, changes in pH, humidity and temperature, microorganisms and food components. However, the results of laboratory studies of water absorption

Table 5.6 Average values of the degrees of water absorption after ten weighings of the samples

Polymer	m_0, g	Δm, g	λ, %	Place in the ranking
Akrodent	3.0073	0.0193	0.642	5
Protemp	3.7072	0.0073	0.197	2
Structur	4.0288	0.0075	0.186	1
Tempron	3.0255	0.0113	0.373	4
Ceramill	3.0080	0.0090	0.299	3

of dental polymers make it possible to predict the term of functional suitability of removable dentures.

5.2.4 Determination of a Shrinkage

Polymers shrinkage is determined during the manufacture of prismatic samples (Type II). To do this, we perform tenfold measurements of their dimensions along and across the axis in the central part, perpendicular to the axis, using a microscope with a hexadecimal magnification. The degree of shrinkage is calculated by the formulas:

by measuring along the axis

$$\Delta l = 100(l_0 - l_1)/l_0, \tag{5.5}$$

where l_0 is the size of the sample after filling the form (mm); l_1—after polymerization (mm);

by measurements perpendicular to the axis

$$\Delta h = 100(h_0 - h_1)/h_0, \tag{5.6}$$

where h_0 is the size of the sample after filling the form (mm); h_1—after polymerization (mm).

Table 5.7 presents the results of measurements with an error of 0.01 mm and the calculated degree of shrinkage of polymers ($l_0 = 39.79$ mm, and $h_0 = 13.6$ mm).

Since the indicator Δl of the shrinkage degree has been obtained for a longer base in the axial direction of measurement, it is more reliable to take into account its value, so the polymers are ranked according to this indicator from the smallest (see Table 5.7).

Table 5.7 Average values of the degree of shrinkage of polymer samples

Polymer	l_1, mm	Δl, %	Place in the ranking	h_1, mm	Δh, %
Akrodent	39.73	0.15	1	12.648	0.07
Protemp	39.41	0.95	4	12.5	0.081
Structur	39.42	0.93	3	12.458	0.084
Tempron	39.41	0.88	2	12.54	0.078
Ceramill	39.46	0.96	5	12.52	0.082

5.2.5 *Microhardness of Polymers*

Microhardness is measured with a device PMT-3 by pressing a diamond pyramid with a load of 50 g (HV_{50}). During the study of each polymer, at least five indentations of the microhardness indenter (impressions) are performed and the dimensions of their diagonals are measured. Microhardness is calculated from the average value of these dimensions. The obtained results are presented in Table 5.8.

For each experimentally established parameter, the polymers are ranked according to a complex indicator, using the sum of places in the rankings given in Tables 5.1, 5.2, 5.5, 5.6, 5.7, and 5.8. The materials are arranged in the following order (from the best): Protemp 4, Akrodent, Structur 2SC, Tempron 1-1PKG, Ceramill PMMA (Fig. 5.10).

The ranking technique involves a large number of experiments, which require significant material and time costs. Therefore, it is important to find such a way of ranking that would reduce costs and save the obtained result.

Table 5.8 Average values of microhardness of polymers

Polymer	HV_{50}, MPa	Place in the ranking
Akrodent	175.2	3
Protemp	226.5	1
Structur	181.1	2
Tempron	139.5	4
Ceramill	140.8	5

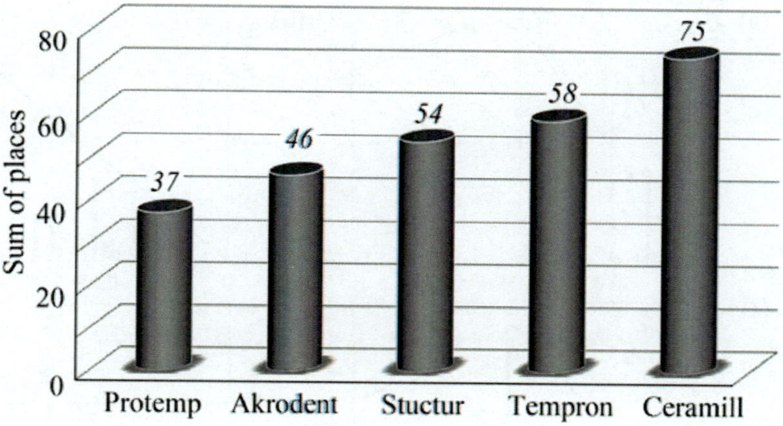

Fig. 5.10 Ranking of polymers by the complex indicator

5.3 Evaluation of the Types of Dental Polymer Fracture by the Energy Criterion

5.3.1 Analysis of AE Under Tensile Fracture of Polymers

Simultaneously with the tension of samples, the AE information is recorded using the detection system SKOP-8M [62]. Its main settings are the same as in previous experiments (see Sect. 5.2).

In the post-operation mode, the stress–strain curves and the distribution of amplitudes (Fig. 5.11) as well as CWT of the detected AE signals are constructed. It is found that all the studied polymers are failed quasi-brittle, which is typical for this class of materials [59].

To evaluate the fracture types in the polymers, CWTs are constructed for AE signals obtained at the initial loading (point A), in the middle section of the curve (point B) and when the critical value of the load (points C and D) is reached. The values of the criteria energy parameter E_{WT} calculated by the constructed method [67] and formula (3.3) are given in Table 5.9.

According to the energy criterion for identifying the types of structural materials fracture by CWT at the initial stages, in the majority of polymers (Tempron, Akrodent, Protemp, Ceramill) AE signals are generated that correspond to the ductile–brittle fracture ($E_{WT} < 0.01$). In the Structur material, at first AE signals correspond to the ductile fracture ($0.01 \leq E_{WT} < 0.1$) Under further loading, the brittle mechanisms prevail, as evidenced by the high values of the AE criterion parameter ($E_{WT} > 0.2$).

The Akrodent polymer is chosen for a more detailed analysis. Figure 5.12, along with a typical curve of the stress–strain and AE activity (Fig. 5.12a), shows the

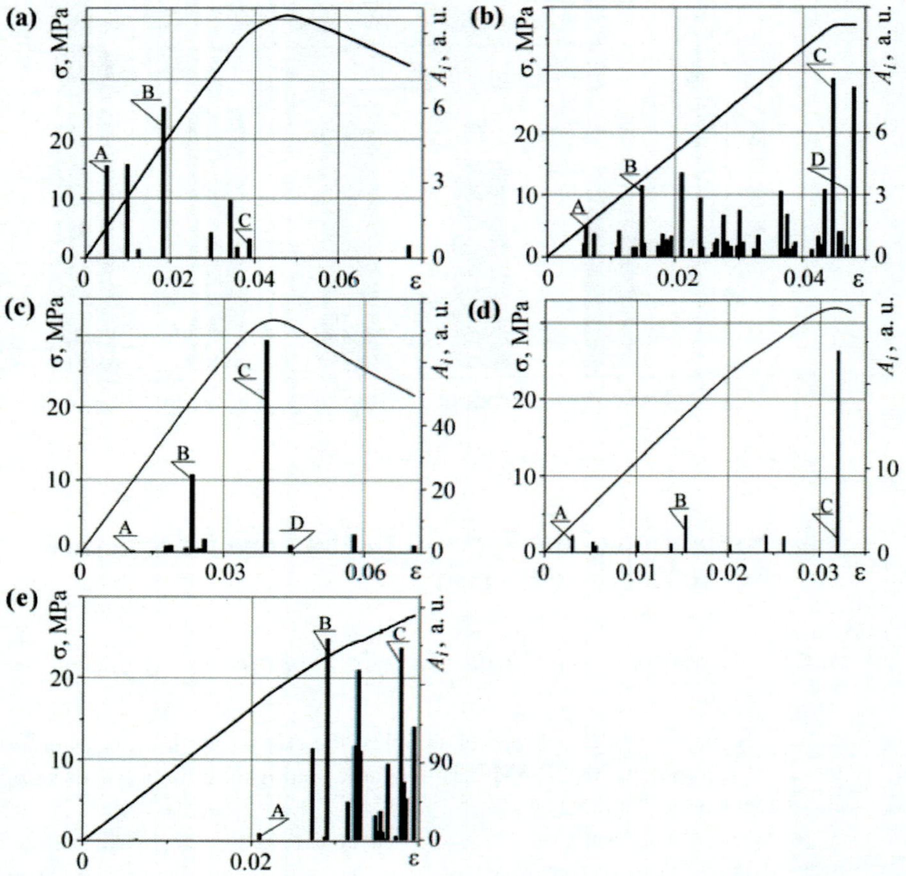

Fig. 5.11 Stress–strain curves and distribution of the amplitudes of AE signals for the samples of polymer materials: **a** Protemp; **b** Akrodent; **c** Structur; **d** Tempron; **e** Ceramill

Table 5.9 Value of the criterion parameter E_{WT} of the corresponding AE signals for different values of polymer samples loading (see Fig. 5.11)

Polymer	E_{WT}			
	A	B	C	D
Structur	0.0047	0.489; 0.206	0.384	0.023
Tempron	0.015	0.13	0.1	–
Akrodent	0.01	0.013	0.1	0.006
Protemp	0.039	0.285	0.012	–
Ceramill	0.019	0.089	0.101	–

distribution of the AE signals amplitudes in an arbitrarily selected interval of 3.5–4.0% of the relative deformation of the sample (Fig. 5.12b), which is drawn by a dashed line in Fig. 5.12a.

As we can see, AE signals of different amplitudes alternate in different sections of the stress–strain curve (Fig. 5.12). Previously [68], this was hypothetically explained by the alternation of the brittle (jump-like crack growth) and ductile (initiation and development of plastic deformations) fracture. The WT of AE signals confirms these hypotheses. The values of energy parameters of local excitations (pulses) calculated by the constructed energy criterion allow us to perform the corresponding identification. Table 5.10 presents its results for the signals generated during tension of the Akrodent polymer sample: points A, B, C, D in Fig. 5.12a and signals *1–6* in Fig. 5.12b.

As follows from Tables 5.9 and 5.10, under tension of the polymer, AE signals alternate, which correspond to different fracture types. This can be interpreted as the

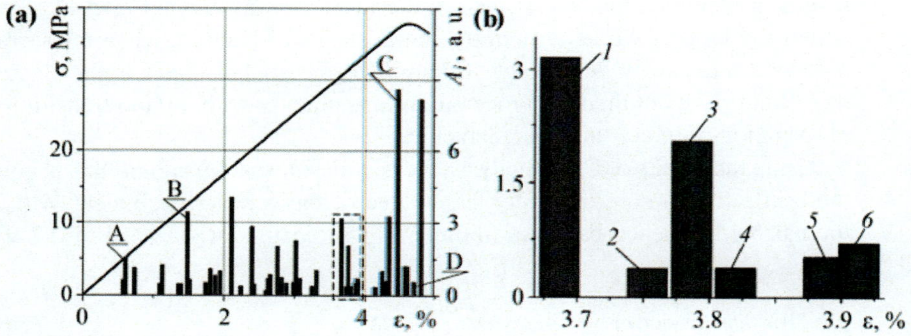

Fig. 5.12 Diagram of the stress–strain and AE activity for the sample of dental polymer Akrodent (**a**) and distribution of the amplitudes of AE signals in the range of the sample relative deformation 3.5–4.0% (**b**)

Table 5.10 Values of energy parameters E_{WT} of local signal pulses and the corresponding fracture type in Akrodent

AE signal (see Fig. 5.12)	E_{WT}	Fracture type
A	0.010	Ductile–brittle
B	0.023	Ductile–brittle
C	0.100	Brittle
D	0.006	Ductile
1	0.100	Brittle
2	0.007	Ductile
3	0.031	Ductile–brittle
4	0.003	Ductile
5	0.003	Ductile
6	0.010	Ductile–brittle

development of the micro- and macrocracking with the formation of plastic zones in front of the crack.

The presence of several local maxima of wavelet coefficients in a short period of time is the peculiarity of the CWT for AE signals, which accompany the fracture of Structur and Ceramill. This may mean the overlapping in time of various fracture events, which according to the values of the criteria parameter correspond to the brittle or ductile–brittle type.

Figure 5.13 presents quantitative indicators of different fracture types during experimental testing of dental polymers, established by κ [59] and energy criteria. It is found that the energy criterion makes it possible to select AE events for all materials that correspond to the plastic deformation, which could not be done earlier by the κ-criterion. This once again confirms the effectiveness of the energy approach to identify the types and, consequently, the mechanisms of materials fracture.

Protemp (16%) has the lowest percentage of brittle fracture, followed by Akrodent (40%), Structur (43%), Tempron (44%), and Ceramill (45%). According to the results of identification and analysis of the stress–strain curves (Fig. 5.11), Protemp shows the highest AE activity in the initial stages of loading, when microcracking has been intensively developing. Ceramill begins to fail under higher loads than other materials, but then the process develops quickly with a rapid transition to the propagation of micro- and macrocracks.

Using the results of CWT analysis, the peculiarities of AE signals that accompany the brittle fracture are relatively large value of the wavelet coefficient WT_{max} and the narrow frequency band Δf in the WT–f projection at the level of 0.7 WT_{max}.

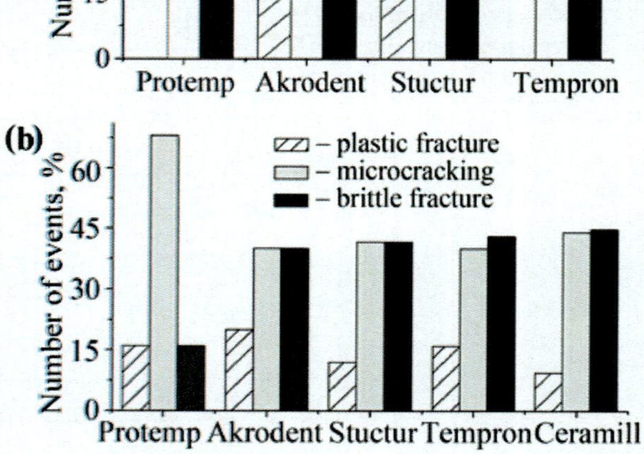

Fig. 5.13 Distribution of local pulses that correspond to different fracture types of dental polymers according to κ (**a**) and energy (**b**) criteria

For the CWT of AE signals that accompany ductile or ductile–brittle fracture, the WT_{max} value is much lower than for brittle fracture, and the corresponding frequency band is significantly wider. The frequency f_{max} corresponding to WT_{max} is less than during ductile or ductile–brittle fracture, because with the defect size increase, the dominant frequency of the corresponding elastic oscillations decreases [69].

Figure 5.14 shows the CWT for the typical AE signals, which correspond to the brittle (Fig. 5.14a) and ductile–brittle (Fig. 5.14b) fracture of the provisional material Protemp. In this case, for AE signal under the brittle fracture we have $WT_{max} = 0.15$, bandwidth $\Delta f = 0.11$ MHz, the duration of radiation $\Delta t = 17.5$ μs, $f_{max} = 312$ kHz, the frequency corresponding to WT_{max} (Fig. 5.14a). For the ductile–brittle fracture, the corresponding parameters are: $WT_{max} = 0.046$, $\Delta f = 0.15$ MHz, $\Delta t = 10$ μs, $f_{max} = 359$ kHz (Fig. 5.14b).

To understand the character of fracture under the action of load in polymer materials, we have studied their microstructure (Fig. 5.15).

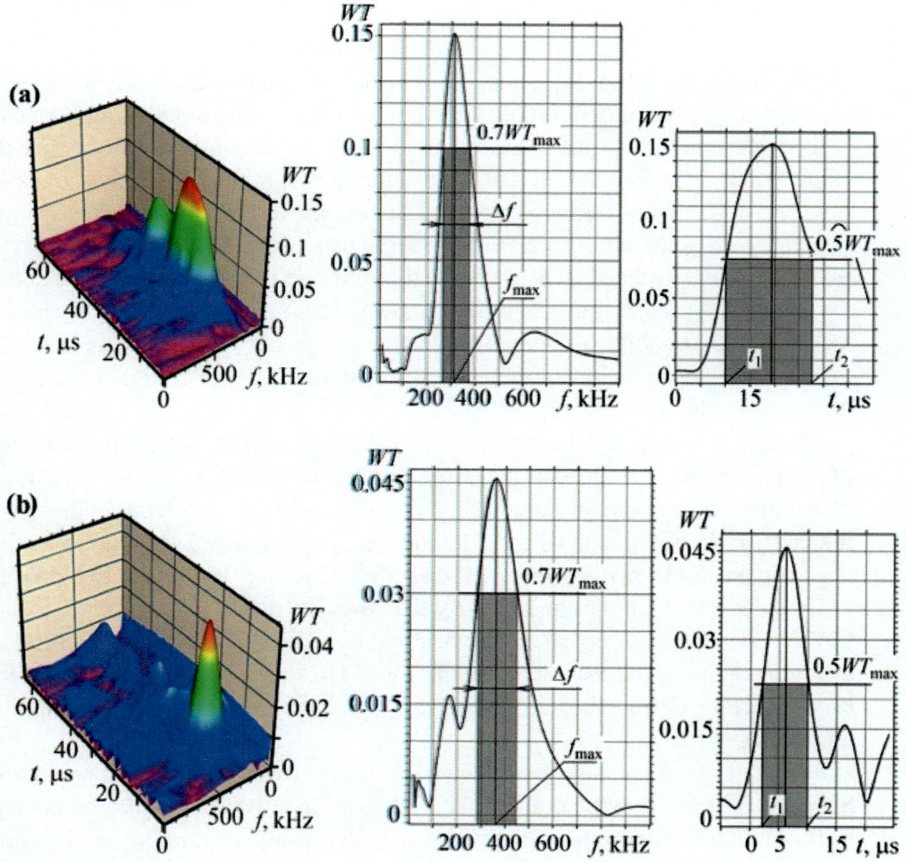

Fig. 5.14 CWT of AE signals for the brittle **a** and ductile–brittle **b** fracture of the Protemp polymer (3D-representation, 2D-time projections of its local impulses at the moment of reaching the value WT_{max})

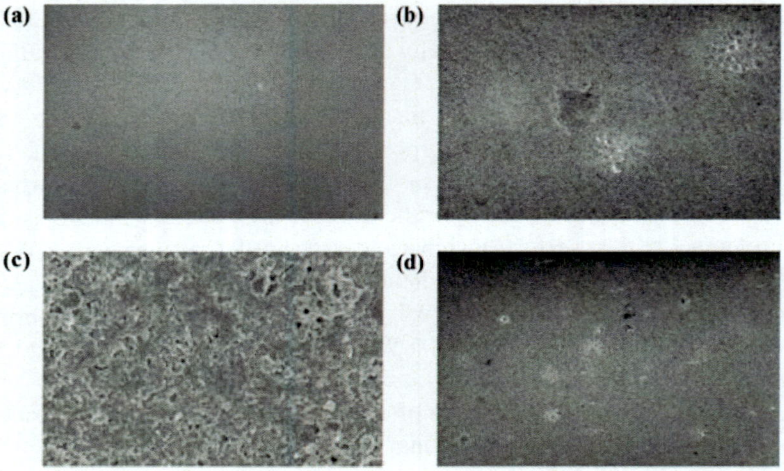

Fig. 5.15 Microstructure of provisional polymers: **a** Protemp; **b** Akrodent; **c** Structur; **d** Tempron

We can see from Fig. 5.15a that Protemp has the highest homogeneity to compare with the other materials. Tempron has a fine structure with a small number of cavities (Fig. 5.15d). The porosity of Akrodent is much larger and is locally concentrated (Fig. 5.15b). For Structur, larger pores alternate with cluster sections of the filler microparticles. Thus, during Akrodent fracture, the largest number of AE signals is generated due to its significant porosity. Protemp has the least of AE activity because of its homogeneity (see Fig. 5.11). Since the pores in Structur are the largest and cause the greatest brittleness, the amplitudes of AE signals and the values of the energy index E_{WT} are the highest. A specific feature of Ceramill is the lowest strength (see Fig. 5.11e), but a significant increase in AE activity occurs at higher loads than in other materials.

Therefore, based on the analysis of fracture diagrams, microstructure of the surface of materials, and the values of the criteria parameter of fracture types E_{WT}, it can be stated that all used dental polymers fracture quasi-brittle. Protemp is the strongest material, and Structur can be considered as the most brittle. Ceramill has the lowest strength, but starts to fracture under loads exceeding this value for other polymers.

To study the dynamics and identify the types of fracture of polymeric materials, CWT of AE signals is considered to be effective. The alternation of AE signals with different values of the criteria parameter E_{WT} indicates the ductile, ductile–brittle, or brittle fracture, despite the fact that the macrofractures of the samples of all polymers are brittle.

Protemp has the smallest percentage of brittle fracture, followed by Akrodent, Structur, Tempron, and Ceramill. The ranking of polymers by the percentage of brittle fracture under tension coincides with the results of ranking by the complex index, as well as by the parameters of strength and AE activity under tension. Based on this, a new express method of ranking dental polymers is proposed. It foresees the experiments under tension until complete fracture of the material samples with simultaneous recording of acoustograms only. By determining the strength of materials, AE activity and establishing the percentage of brittle fracture by the energy criterion for identifying its types, one can significantly reduce the experimental and time costs for ranking polymers, at the same time ensuring the correspondence of the results to the complex indicator.

5.4 Peculiarities of Some Tooth-Endocrown Systems Fracture Under Quasi-Static Loading

According to the analysis of literature, the AE activity, amplitude and total energy of AE signals, signal duration or its rise time, maximum frequency, and so on are the informative parameters of AE process. Recently, during AE signals processing, WT has been widely used, which makes it possible not only to select the frequency components of the signal, but also to localize them in time. Such local analysis of AE signals can be effective both for assessing the kinetics of fracture development and for determining the initiation of fracture of the components of orthopedic constructions, because the initiation of cracking is clinically most important.

5.4.1 Preparation of the Samples

Five dental materials have been studied [70]. The samples are made according to standard technologies in the form of disks 5 mm thick and 13 mm in diameter (five pieces of each material). They are divided into five types: B—zirconium dioxide (Prettau zirconia), C—ceramics (IPS e.max Press), D—metal ceramics (GCInitialMC), E—composite resin (FineNanoQ), F—luting cement (RelyX™ U200) (Table 5.11).

Twenty-five human maxillary and mandibular molars extracted not earlier than two months prior to starting the experiment have been selected for testing. Teeth have been cleaned mechanically, using a synthetic brush and ultrasonic scaler of the dental unit and chemically during exposure to 100 ml of 6% sodium hypochlorite solution (Xloraxid 6% Cerkamed), heated to 60 °C, for 20 min. They are then divided into five types (five samples each). Teeth without restoration (control) correspond to type A. Teeth restored with an endocrown made of zirconium dioxide (Prettau zirconia) are classified as type BE, of ceramics (IPS e.max Press)—as type CE, of metal ceramics (GC Initial MC)—as type DE and of composite material (FineNanoQ)—as type EE. MIHM-VOGT HT sintering technology is used to make CE-type endocrowns, and Zirconzahn Modellier (Zirconzahn GmbH, Österreich) standard technology and sintering in Zirconzahn 600 furnace are used for BE- and DE-type endocrowns. The DE-type samples are sintered at a temperature of 1500–1600 °C for 9 h. Vita ceramic mass 2 mm thick is applied to the metal frames and sintered in the furnace.

Table 5.11 Brands, types, main chemical compositions, and manufacturers of the materials used for the experiments

Brand	Type	Chemical composition	Manufacturer
IPS e.max Press	Reliable lithium disilicate Glass Ceramics	97% SiO_2, Al_2O_3, P_2O_5, K_2O, Na_2O, CaO, F, 3% TiO_2, pigments, water, ethanol	Ivoclar Vivadent
Prettau zirconia	Zirconium dioxide	ZrO_2, 4–6% Y_2O_3, $Al_2O_3 < 1\%$, SiO_2 max 0.02%, Fe_2O_3 max 0.01%, Na_2O max 0.04%	Zirkonzahn Worldwide
(1) GC0 Initial MC (ceramics) (2) Nicrallium N2 BCS (metal)	(1) Metal ceramics (2) Nickel-Chrome ceramic alloy	(1) 55–65% SiO_2, 10–15% Al_2O_3, 10–15% K_{20}, 5–8% Na_2O, $LiO_2 < 1\%$, $MgO < 1\%$, 1–2% CaO, 0–2% BaO, 0,5–2% B_2O_3, $TiO_2 < 1\%$, $ZrO_2 < 1\%$, $P_2O_5 < 1\%$, $(CeO_2, CeF_3) < 1\%$ (2) 62% Ni, 24% Cr, 10% Mo, 2% Si, 1% Ce, 1% Fe, C max 0.03%	(1) GC EUROPE N.V (2) BERNARD CERVOS SOCIETE
Nano Q	Photo-polymerized resin	Monomer matrix: diurethane dimethacrylate, butanediol dimethacrylate, isopropylide-bis [2(3)-hydroxy-3(2)-(4-phenoxy) propyl] bismethacrylate [Bis-GMA] Total filler content 83% by weight (70% by volume)—Glass filler (medium grit size 0.7 μm)—Pyrogenic silica (medium grit size 12 nm)—Agglomerated nanoparticles (medium grit size 0.6 μm)	GDF Private Label Chemistry
RelyX™ U200	Dual polymerized resin cement	Contains bifunctional (meth) acrylate. The proportion of inorganic fillers is about 43% by volume; the grain size (D 90%) is about 12.5 μm	3 M ESPE

The EE-type samples are coated with BiscoTescera varnish and polymerized by standard technology. All the crowns are fixed in the teeth with RelyxU200 composite cement according to the manufacturer's instructions. Restored and unrestored molars are fixed in a dental low-melting alloy Melot.

Finally, all the samples for research are divided into two groups: 1—samples of B, C, D, E, F materials; 2—orthopedic tooth-endocrown restorations of different materials (types A, BE, CE, DE, EE).

The samples are loaded under compression on the SVR-5 testing machine using a hard steel ball (the ball diameter—4 mm) with a crosshead speed of 0.12 mm/min. The scheme of the test setup is shown in Fig. 5.16. Sample *6* is mounted on a support *1*. A waveguide *2* and a special anti-friction gasket *3* are placed between the sample and the support. An AE sensor *4* is attached to the waveguide with silicone

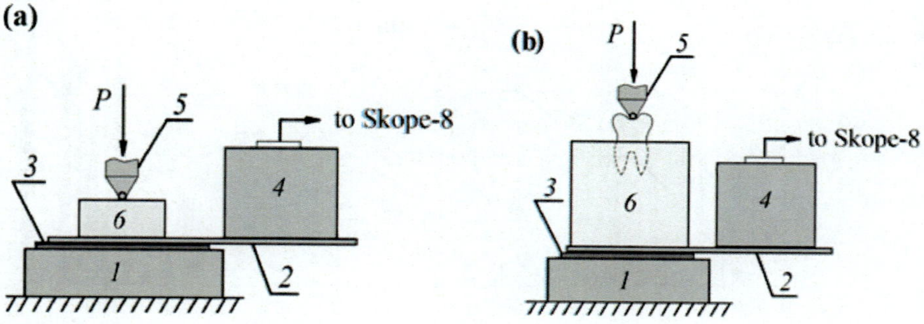

Fig. 5.16 Structure of compression tests: **a** samples of group 1; **b** group 2

grease in order to provide a good acoustic coupling between the waveguide and the piezoelectric sensor. Load is applied to the samples through the indenter *5*.

During the quasi-static compression experiments, AE information is recorded synchronously with loading using the SKOP-8 measuring system [62]. Its main settings are the same as in previous experiments (see Sect. 5.2).

The means and standard deviations of the initial fracture load, which coincide with the beginning of AE activity, and final fracture load are established. All the data are statistically analyzed in the ANOVA program.

To study the specific features of AE signals during fracture of the orthopedic tooth-endocrown restorations, a previously constructed method for estimating the types of fracture by the energy criterion of the CWT of AE signals has been used.

5.4.2 The Initiation of Fracture in the Fixed Orthopedic Tooth-Endocrown Restorations

Figure 5.17 presents the typical graphs of time variation of loading and AE activity under the compression tests of teeth and samples of dental materials.

Note that the load of the zirconium dioxide sample has been limited to the maximum possible value $P = 10$ kN for the experimental setup. Since the AE signals have not been detected, the failure load is not determined.

Table 5.12 shows the means and standard deviations of the initial P_{if} (the initiation of fracture propagation set by the AE signals) and final P_{ff} fracture load for all the materials. Figure 5.18 demonstrates the diagram of distribution of these indicators.

It has been found that in all the samples under their quasi-static compression the fracture initiates at the loads P_{if} much lower than the final fracture load P_{ff}, with the largest difference for ceramic and metal ceramic samples (6.196 and 3.314 kN, respectively) and the smallest one for luting cement (1.26 kN). A similar trend has been obtained earlier in [71] and [72], investigating other dental materials and

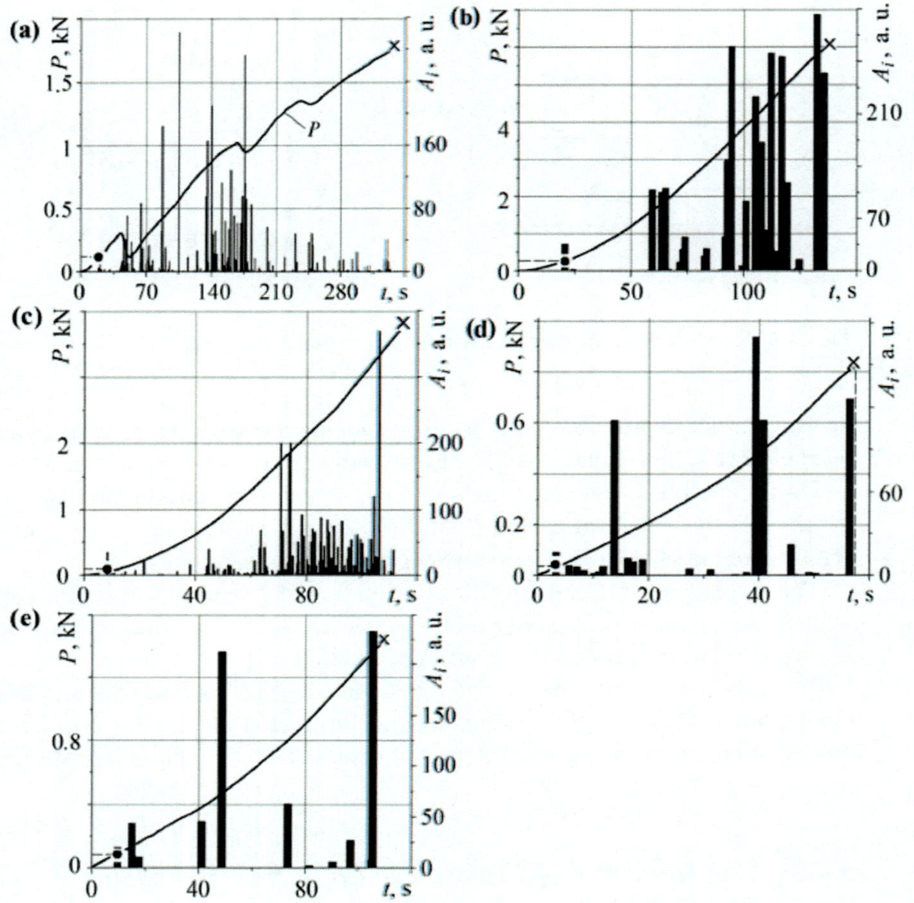

Fig. 5.17 Typical graphs of load *P* and AE activity A_i versus time for the samples of group 1: **a** A; **b** C; **c** D; **d** E; **e** F; ●—initiation of fracture, × – complete fracture

Table 5.12 Means and standard deviations (std) for the initial fracture load P_{if} and the final fracture load P_{ff} for dental materials under their quasi-static compression

Experimental group 1	P_{if} (kN): (±std)	P_{ff} (kN): (±std)
A	0.066 (±0.031)	1.645 (±0.082)
C	0.253 (±0.064)	6.449 (±0.668)
D	0.056 (±0.022)	3.362 (±0.524)
E	0.039 (±0.019)	1.735 (±0.300)
F	0.029 (±0.017)	1.289 (±0.308)

constructions. The determined values of the load at the beginning of fracture development in the ceramics and metal ceramics agree well with those similar to amalgam [73].

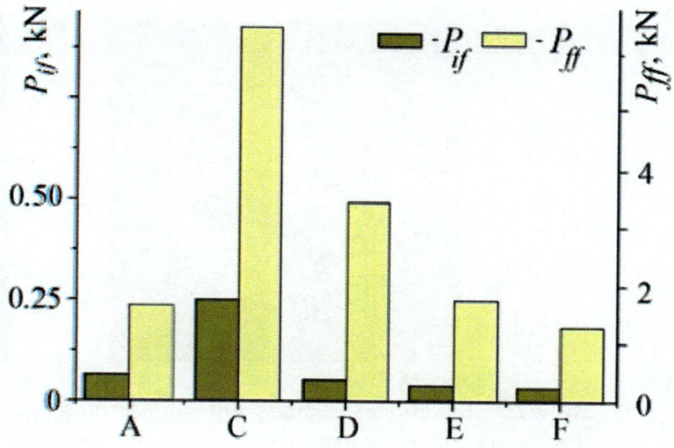

Fig. 5.18 Initial P_{if} and final P_{ff} fracture load for dental materials of group

As can be seen from Fig. 5.18, ceramics which fracture under compressive load has the highest strength among the dental materials, and the lowest has a luting cement. The final fracture load of a tooth (type A) is lower than that of the restoration materials.

Figure 5.19 illustrates the macrofracture of the dental material samples. The ceramic samples fail mainly into several almost identical parts with almost smooth surfaces parallel to the axis of the load application (Fig. 5.19a). The peculiarity of fracture of the samples of composite resin and luting cement is their division into several pieces of different sizes with the crack propagation in different (parallel and non-parallel) directions relative to the axis of the load application (Fig. 5.19b). Similar types of macrofracture of dental materials (amalgam and ceramic-reinforced glass ionomer cement) are established in [73].

Since the zirconium dioxide samples do not crack under the load that approximated 10 kN, this material is ranked top of the list of the dental material under study for strength, followed by ceramics, metal ceramics, tooth, composite resin, and luting cement.

Analysis of the dynamics of the fracture development in the orthopedic tooth-endocrown restorations under compression and photographs of fractured samples of

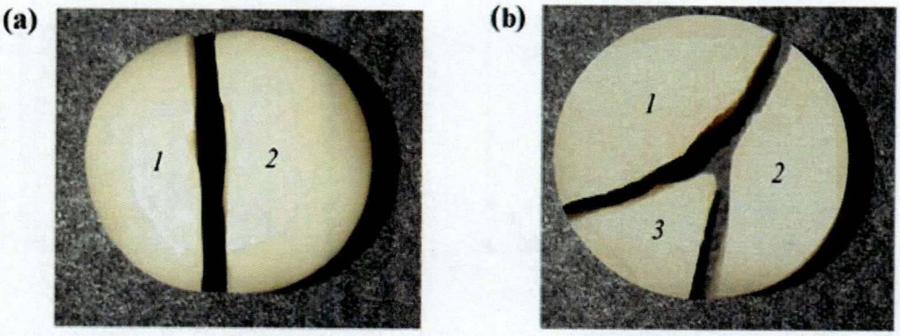

Fig. 5.19 Macroscopic fracture of the samples of dental materials: **a** ceramics, **b** composite resin

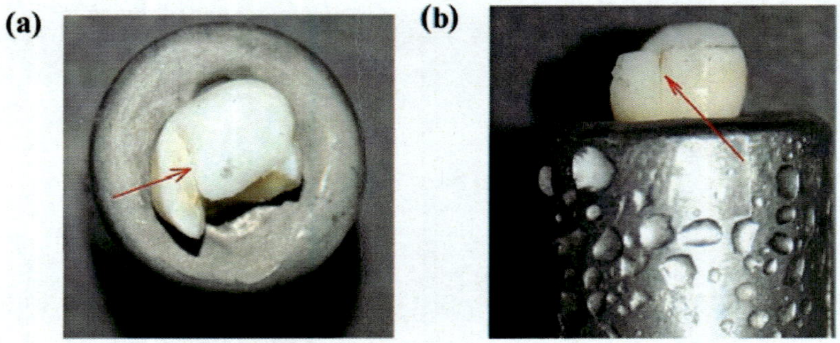

Fig. 5.20 Representative fractured samples of tooth-endocrown restoration with a zirconium dioxide: **a** general view from above, **b** cracks propagation in the tooth (arrowed lines)

the restorations indicate its initiation in the tooth. Figure 5.20 illustrates the fracture of the tooth-endocrown restoration made of zirconium dioxide.

As it is mentioned above, the zirconium dioxide sample has not fractured under the compressive stress, but because of the crack propagation in the tooth. The endocrown made of it, having lost the rigidity of the base, has eventually fractured (Fig. 5.20a).

The strength of the endocrown material also decreases under loading of restorations with the ceramic and composite resin endocrowns. In case of the restoration with the metal ceramic endocrown, some specific features are found (Fig. 5.21). Although its strength has not decreased, the character of the fracture curve changes: During compression, it is similar to such of plastic materials.

This can be explained by the fact that the metal base of the metal ceramic endocrown has obviously increased the fracture toughness due to its elastic characteristics (Fig. 5.22a). The loss of the base rigidity (tooth fracture) is smoothly compensated by the flexibility of the metal cap.

Table 5.13 shows the value of the initial P_{if} and the final P_{ff} fracture load in restorations with endocrowns of different materials, and Fig. 5.23—the corresponding diagram according to its data.

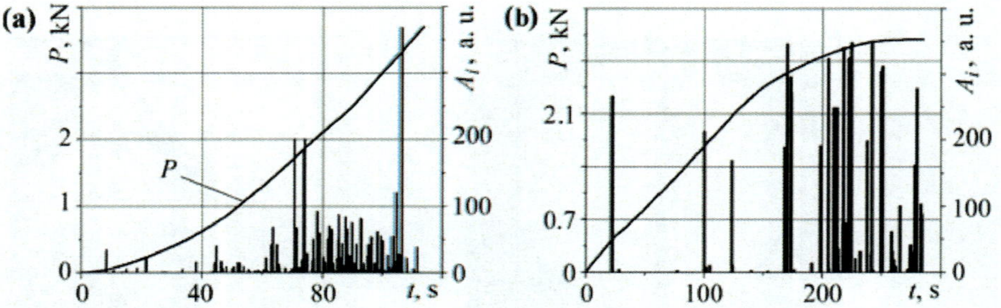

Fig. 5.21 Time dependence of load P during the compression test of a typical cermet sample (**a**) and a restoration with an endocrown made of it (**b**)

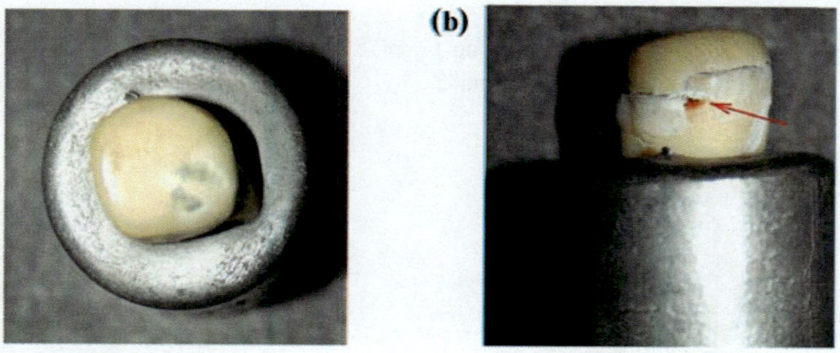

Fig. 5.22 Representative fractured samples of tooth-endocrown restoration with the metal ceramic endocrown: **a** general view from above, **b** cracks propagation in different areas of the sample (arrowed line)

Table 5.13 Means and standard deviations (std) of the initial P_{if} and final P_{ff} fracture load of the tooth-endocrown restorations with different endocrowns

Experimental group 2	P_{if} (kN): (±std)	P_{ff} (kN): (±std)
BE	2.511 (±0.333)	3.082 (±0.305)
CE	0.148 (±0.020)	2.726 (±0.226)
DE	0.072 (±0.031)	3.320 (±0.423)
EE	0.069 (±0.038)	1.533 (±0.211)

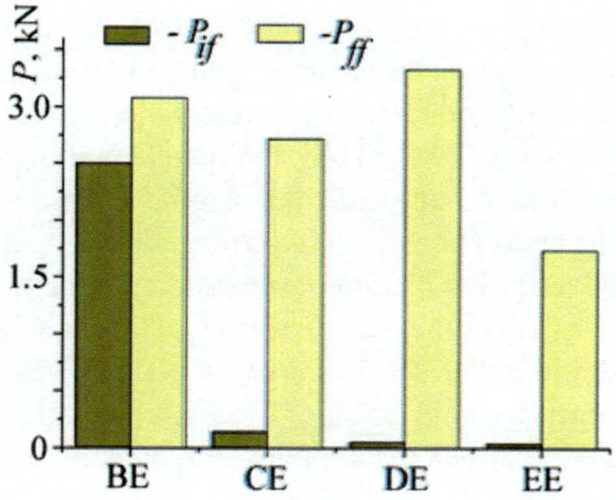

Fig. 5.23 Initial P_{if} and final P_{ff} fracture load for endocrown restorations (group 2)

It has been found (Fig. 5.23) that in restorations with the zirconium dioxide endocrown, a fracture occurs at the highest load compared to the other types of restorations, but the highest values P_{ff} are recorded in the restorations with the metal ceramic endocrown.

Thus, in all the samples of orthopedic restorations under their compression, the fracture starts at the lower loads P_{if}, than the maximum P_{ff}. At the same time, for restorations with the metal ceramic endocrown, the difference between the initial P_{if} and final P_{ff} fracture load is much larger than for others. This is probably due

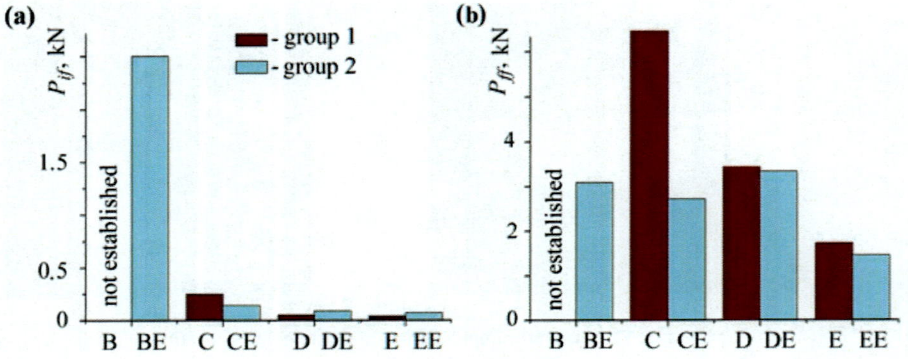

Fig. 5.24 Initial P_{if} (**a**) and final P_{ff} (**b**) fracture loads for dental materials and for endocrown restorations with endocrowns made of the respective materials

to the presence of a metal cap, which contributes to the plasticity of the restoration and restrains its complete fracture under increasing deformation due to the loss of rigidity of the base (tooth).

Figure 5.24 illustrates the diagrams of loads P_{if} and P_{ff} in dental materials under their compression separately and in the tooth-endocrown restoration.

It is found (Fig. 5.24a) that during compression the load P_{if}, which initiates the fracture in the ceramic samples, is slightly higher than in the restorations with endocrowns of such material (0.25 and 0.148 kN, respectively). At the same time, for the samples made of metal ceramics and composite resin, on the contrary, it is slightly lower than under compression of restorations with endocrowns made of these materials (0.056 and 0.072 kN; 0.039 and 0.069 kN, respectively).

The maximum load P_{ff} of the dental material samples (Fig. 5.24b) in all cases is higher than of the restorations with the corresponding endocrowns. For the restorations with the endocrowns of zirconium dioxide and ceramics, the fracture load (3.082 and 2.726 kN, respectively) is significantly less than that of the endocrown material (not established and 6.449 kN, respectively). This phenomenon can be explained by the fact that during compression of the restoration, the intensive development of deformations and fracture in the tooth leads to the loss of rigidity of the base (tooth) of the endocrown and increases the deflection of the material, and thus accelerates the endocrown fracture. That is, the fracture resistance of structures with endocrowns is definitely influenced by the deflection of the endocrown material. As a result, the failure of the orthopedic restoration in general is accelerated. Therefore, the strength of the restoration depends on the strength of its base, namely the tooth.

Thus, the orthopedic tooth-endocrown restoration is a complex mechanical system, the initiation of fracture of which depends on many physical factors. In particular, this phenomenon depends on changes under loading of the base (tooth) geometry, its biological state, and other physiological features. The initiation of fracture of such a system starts with the tooth. The restoration with an endocrown made of metal ceramics has the lowest risk of premature failure during a clinical use.

5.4.3 Specific Features of AE Signals Under Fracture of the Tooth-Endocrown Restorations. Analysis of AE Signals During Tooth Fracture

According to the spectral distribution (FFT) of AE signals obtained in the tooth samples fracture, their two types are distinguished: I—with an amplitude peak in the band of 120–130 kHz; I—with peaks in the bands of 120–130 and 240–350 kHz, where the amplitudes in the first frequency band are dominating.

According to the above-presented algorithm for estimating the local maxima of AE signals, their parameters are determined and the type of failure is established by the energy criterion. The signals of the initial stages of fracture, denoted by numbers *1–7*, are considered (Fig. 5.25a). It is found that in the time range of 45–50 s from the beginning of the experiment there is a sharp drop in the load, which, apparently, corresponds to the intensive development of tooth fracture. For a detailed analysis of the signals of this time interval, similar calculations are performed for the corresponding AE signals from the interval between points A and B (signals *1*–7** in Fig. 5.25b). Signals *1–5, 1*–4** are classified as Type I. To the Type II belong the signals *6, 7, 5*–7**. According to the results of the identification of fracture types, it is found that signals *1, 4, 5, 1*–4** correspond to the brittle fracture and the signals *2, 3* show the microcracking. Signals with two peaks (*7, 5*–7**) are mainly generated (according to our identification method) by microcracking and crack propagation, and in signal *6* one of the local pulses characterizes a slight plastic deformation.

Therefore, the early stages of tooth fracture are accompanied by Type I signals, which alternate with Type II AE signals. According to the proposed energy criterion for identifying types of fracture [73], Type I signals correspond to the brittle fracture (occurring due to the formation of new fracture surfaces of a large area), and Type II—to ductile–brittle (caused by the initiation and subsequent growth of microcracks).

Figure 5.26 presents the dependences of the energy parameter of local pulses of AE signals on the maximum frequency, bandwidth, and duration of their radiation. Local pulses of CWT of AE signals, which characterize brittle fracture, have a narrow frequency band Δf (40–75 kHz) (Fig. 5.26a), a long duration of radiation

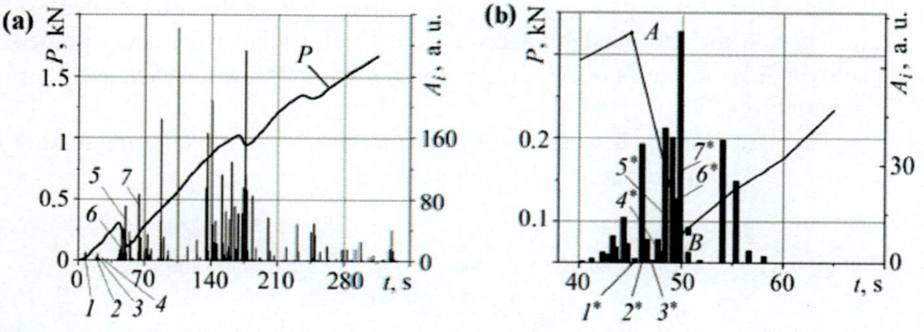

Fig. 5.25 Change of load during the experiment (**a**) and in the interval 45–50 s from its start (**b**)

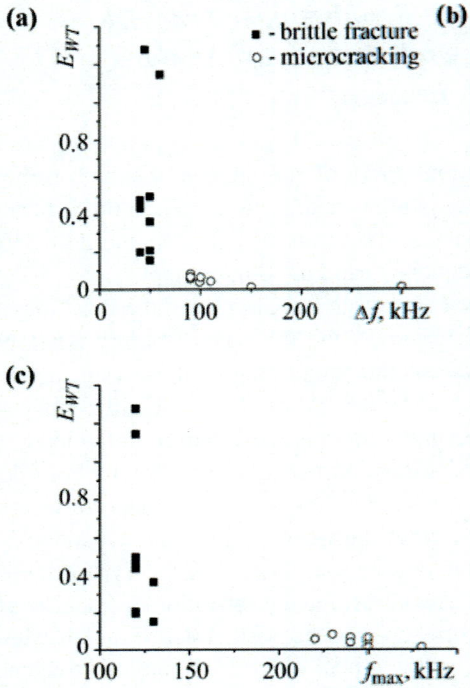

Fig. 5.26 Energy parameter of local pulses of AE signals during tooth fracture depending on: **a** the bandwidth Δf; **b** duration of their radiation Δt; **c** maximum frequency f_{max}

Δt (27–70 μs) (Fig. 5.26b) and large values of energy parameters E_{WT} (0.155–1.276) (Fig. 5.26c) to compare with local pulses, which characterize microcracking. In this case, the ranges of change of these parameters values are: Δf—50–300 kHz, Δt—10–24 μs, E_{WT}—0.013–0.083, respectively.

One more typical feature of AE signals corresponding to the brittle fracture is low values of frequencies f_{max}, which correspond to the values of wavelet coefficients WT_{max} of local pulses, compared with the similar parameters of AE signals during microcracking. Thus, with a decrease in the dominant frequency under brittle fracture, the energy parameters of local pulses and the duration of their radiation increase, while the bandwidth decreases. Similar regularities are recorded when studying elastic waves of AE due to the increase in the defect size in the structural material [69, 74].

Thus, the dominant mechanism of the tooth fracture under compression is the macrocrack propagation, i.e., a brittle fracture.

5.4.4 Evaluation of AE Signals Under Fracture of the Samples Made of Different Materials

According to the spectral distribution of AE signals obtained under fracture of the *ceramic* samples, their two types are distinguished: I—with an amplitude peak in the band of 135–140 kHz; II—with the distribution of peaks in the bands of 135–140 and 250–260 kHz. The frequency band of 135–140 kHz prevails. According to the energy criterion, all AE signals at the beginning of fracture of the ceramic samples characterize the brittle fracture (Fig. 5.27); i.e., with increasing stress in the sample, obviously, new defects are formed with an area commensurable with the macrocrack size.

In accordance with the spectral distribution of AE signals obtained during fracture of the *metal ceramic* samples, their three types are distinguished: I—with one amplitude peak in the band of 120–125 kHz; II—with peaks in the bands of 120–125 and 240–250 kHz; III—with peak distribution in the bands of 120–125 and 420–500 kHz. The emergence of peaks in the higher frequency band is probably caused by the peculiarity of the metal ceramic samples structure. The metal base contributes to the fact that significant elastic deformations can occur in the sample under the load action.

At the same time, considering the results presented in Fig. 5.27, it can be argued that during compression loading of metal ceramic samples, the fracture occurs due to intense microcracking and subsequent propagation of the macrocracks (brittle fracture).

By the spectral distribution of AE signals obtained under fracture of *composite resin* samples, it is found that their specific feature is the distribution of amplitude peaks in the bands of 150–160 (dominant band) and 240–250 kHz. Its fracture has started almost immediately after the load application and is characterized by the alternation of the mechanisms of microcracking and brittle fracture (Fig. 5.27).

According to the spectral distribution of AE signals obtained under fracture of the *luting cement* samples, two types are distinguished: I—with the unique amplitude

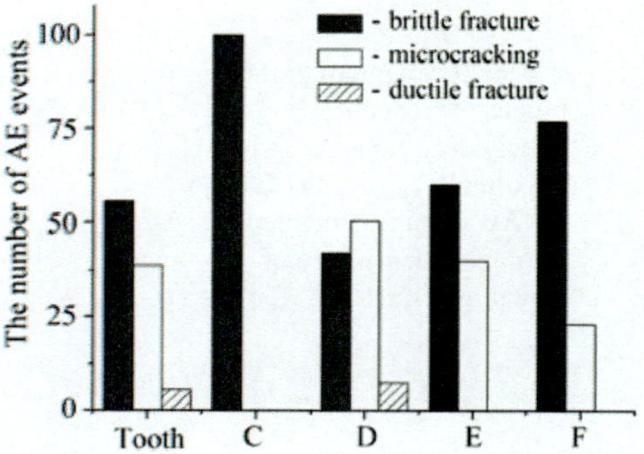

Fig. 5.27 Distribution of local pulses corresponding to different types of fracture (in percent) of dental materials (group 1)

peak in the band of 120–130 kHz; II—with peaks in the bands of 120–130 and 240–250 kHz. The frequency band of 120–130 kHz dominates. Brittle fracture prevails (Fig. 5.27).

The samples of zirconium dioxide have not fractured under compression load.

Figure 5.28 presents a generalized diagram of the dominant frequency bands of AE signals that accompany fracture of dental materials. As one see, the bands obtained by their spectral distribution are different. The AE signals, which accompany the metal ceramic samples fracture, have the narrowest frequency band. Although the frequency bands partially overlap during the fracture of a tooth, metal ceramics, and luting cement, they can be distinguished by the average values of the maximum wavelet coefficients of CWT for the corresponding AE signals. During the fracture of ceramics and composite resin, the slightly higher frequencies dominate in AE signals than under the fracture of a tooth, metal ceramics, and luting cement.

In Fig. 5.29, a diagram of the ranges of the bandwidth values Δf of the local maxima WT_{max} of CWT of the AE signals, generated during brittle fracture of dental materials, is constructed.

So, the largest range of the frequency bandwidth values of local maxima WT_{max} have the AE signals, generated under brittle fracture of the tooth. This indicates the propagation of cracks of different sizes under compression loading—from micro-

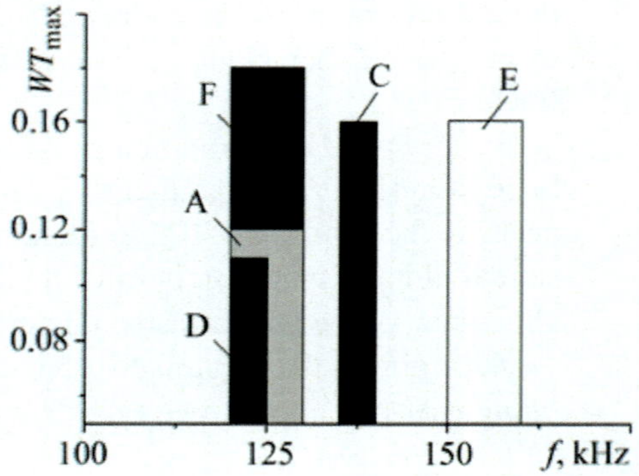

Fig. 5.28 Diagram of the bands of dominant AE signals frequencies that accompany the tooth and dental materials (group 1) fracture

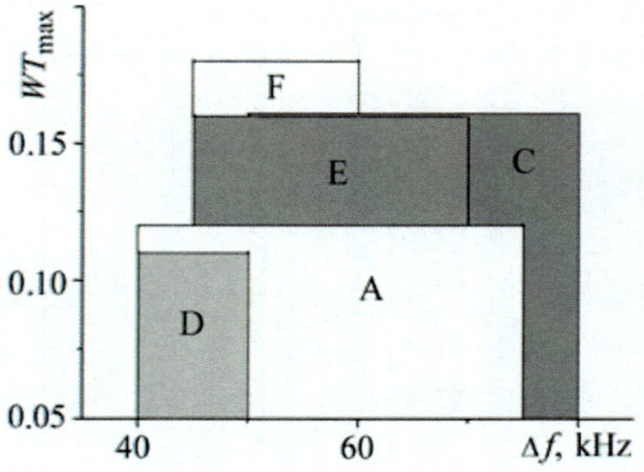

Fig. 5.29 Diagram of the ranges of frequency bandwidth values Δf of local maxima WT_{max} of the CWT for AE signals, generated during brittle fracture of dental materials

to macrocracks, because it is known that the bandwidth depends on the size of a newly formed defect [69]. The smallest range of the bandwidth values of local maxima WT_{max} have the AE signals generated during compression of metal ceramics. This mean that the brittle fracture in it propagates due to the formation of smaller (compared to other materials) cracks of approximately the same size. It is found (Fig. 5.29) that although the frequency bands overlap, the amplitude indicators of AE signals for different materials are differ.

The obtained results correlate well with the results presented in [73], where the types of fracture of dental restoration materials (amalgam and ceramic-reinforced cement) are determined on the basis of microscopic studies of their fracture surfaces. The authors have identified three types of fracture: radial crack propagation; conical cracking, which starts from the load zone; subsurface plastic deformation.

5.4.5 *Analysis of AE Signals During Fracture of the Tooth-Endocrown Restorations*

As evidenced by the analysis of the spectral distribution of AE signals generated during fracture of orthopedic restorations of different types (with endocrowns of zirconium dioxide, ceramics, metal ceramics and composite), their dominant frequency bands are the same—240–250 and 400–450 kHz. According to the energy criterion, the mechanism of brittle fracture prevails during the fracture of restorations of different types at the initial stages of loading (Fig. 5.30).

The ranges of values of the frequency bandwidth of the local maxima WT of AE signals during brittle fracture (Fig. 5.31) are almost the same for all the types of constructions. This is additional confirmation that the fracture of restorations under compression is brittle, although, as can be seen from Fig. 5.30, the bands of the dominant frequencies for different materials vary from each other.

It is found (Fig. 5.32) that the fracture in orthopedic restorations under the action of axial compression load starts with the tooth.

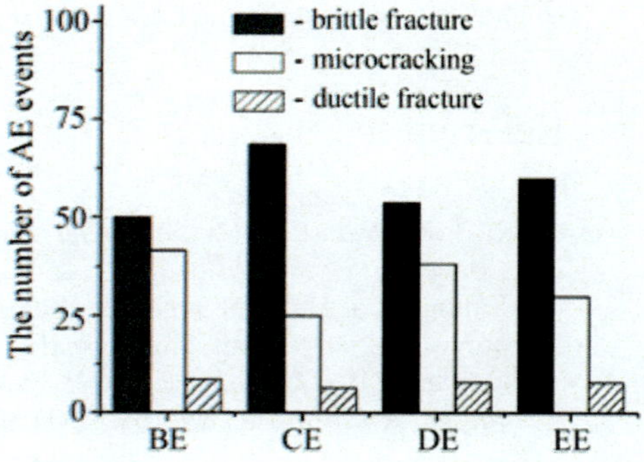

Fig. 5.30 Distribution of local pulses corresponding to different types of fracture (in percent) in the tooth-endocrown restorations (group 2)

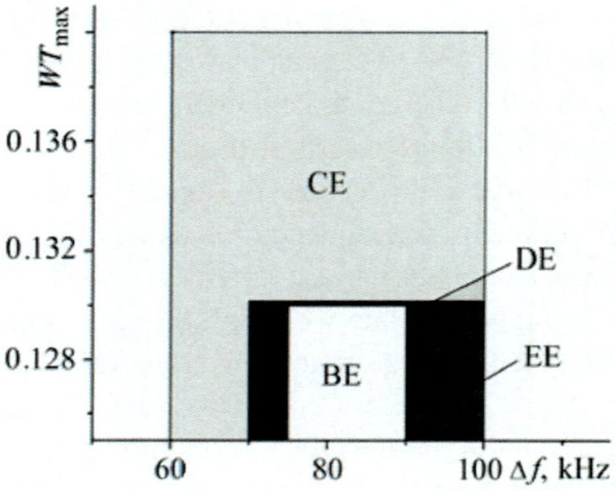

Fig. 5.31 Diagram of the ranges of frequency bandwidth values Δf of local maxima WT of AE signals, generated under brittle fracture of orthopedic restorations of different types (group 2)

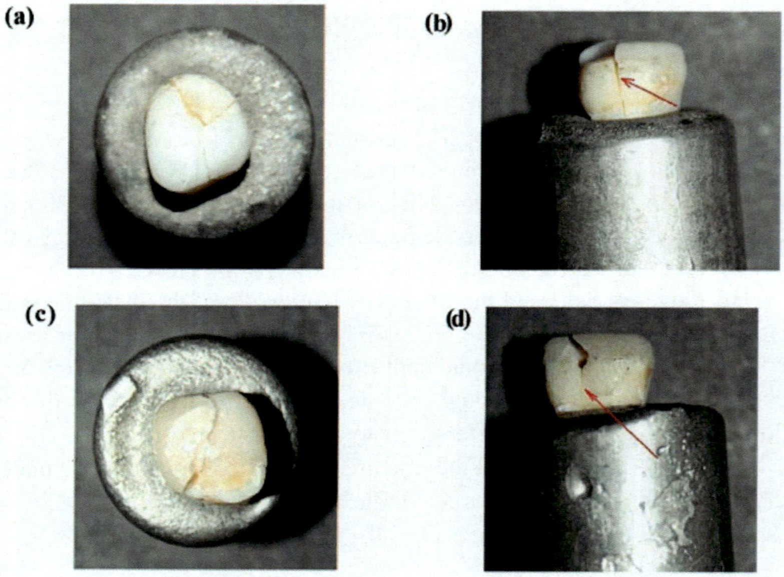

Fig. 5.32 Photograph of the fractured restorations with an endocrown made of ceramics (**a**, **b**) and composite (**c**, **d**): **a**, **c** general top view; **b**, **d** propagation of cracks in the restoration

References

1. O'Brien, W. J. (Ed.). (2008). *Dental materials and their selection* (4th ed.). Quintessence Publ Co Inc.
2. Koumjian, J. H., & Nimmo, A. (1990). Evaluation of fracture resistance of resin used for provisional restorations. *The Journal of Prosthetic Dentistry, 64*, 654–657.
3. Osman, Y. I., & Owen, C. P. (1993). Flexural strength of provisional restorative materials. *The Journal of Prosthetic Dentistry, 70*(1), 94–96.

4. Ireland, M. F., Dixon, D. L., Breeding, L. C., & Ramp, M. H. (1998). In vitro mechanical property comparison of four resins used for fabrication of provisional fixed restorations. *The Journal of Prosthetic Dentistry, 80*, 158–162.
5. Haselton, D. R., Diaz-Arnold, A. M., & Vargas, M. A. (2002). Flexural strength of provisional crown and fixed partial denture resins. *The Journal of Prosthetic Dentistry, 87*(2), 225–228.
6. Lang, R., Rosentritt, M., Behr, M., & Handel, H. (2003). Fracture resistance of PMMA and resin matrix composite-based interim FPD materials. *The International Journal of Prosthodontics, 16*(4), 381–384.
7. Rosentritt, M., Behr, M., Lang, R., & Handel, H. (2004). Flexural properties of prosthetic provisional polymers. *The European Journal of Prosthodontics Restorative Dentistry, 12*(2), 75–79.
8. Balkenhol, M., Mautner, M. C., Ferger, P., & Wöstmann B. (2008). Mechanical properties of provisional crown and bridge materials: chemical-curing versus dual-curing systems. *Journal of Dentistry, 36*(l), 15–20.
9. Kim, S. H., & Watts, D. C. (2007). In vitro study of edge-strength of provisional polymer-based crown and fixed partial denture materials. *Dental Materials, 23*(12), 1570–1573.
10. Scherrer, S. S., Wiskott, A. H. W., Coto-Hunziker, V., & Belser, U. C. (2003). Monotonic flexure and fatigue strength of composites for provisional and definitive restorations. *The Journal of Prosthetic Dentistry, 89*(6), 579–588.
11. Babcic, A., Perry, R., & Kugel, G., *Selected mechanical properties of temporary crown and bridge.* Retrieved 30 March, 2021, from https://multimedia.3m.com/mws/media/597787O/protemp4-protempcrown-clinicalresults.pdf
12. Arutyunov, S. D., & Chumachenko, E. N. (2005). Analiz prochnostnykh kharakteristik konstrukcyonnoho materiala "Akrodent", ispolzuyemoho v technologii provizornych protezov (Analysis of the strength characteristics of the structural material "Acrodent" used in the technology of provisional dentures). *Panorama ortopedicheskoi stomatologii (Panoramic view of orthopedic dentistry), 4*, 34–37.
13. Polonsky, R., Penkner, K., Wender, S., Haas, M, & Bratschko, R. O. (1998). Belastbarkeit moderner provisorienkunststoffe. *Deutsche Zahnärztliche Zeitschrift, 53*(11), 795–799.
14. Frankenberger, R., Pashley, D. H., Reich, S. M., Lohbauer, U., Petschelt, A., & Tay, F. R. (2005). Characterisation of resin-dentine interfaces by compressive cyclic loading. *Biomaterials, 26*(14), 2043–2052.
15. Lohbauer, U., Von der Host, T., Frankenberger, R., Krämer, N., & Retschelt, A. (2003). Flexural fatigue behavior of resin composite dental restoratives. *Dental Materials, 19*(5), 435–440.
16. Yap, A. U. J., Low, J. S., & Ong, L. F. (2000). Effect of food-simulating liquids on surface characteristics of composite and polyacid-modified composite restoratives. *Operative Dentistry, 25*(3), 170–176.
17. Yap, A. U. J., Tan, D. T., Goh, B. K., Kuah, H. G., & Goh, M. (2000). Effect of food-simulating liquids on the flexural strength of composite and polyacid-modified composite restoratives. *Operative Dentistry, 25*(3), 202–208.
18. Yap, A. U. J., Mah, M. K. S., Lye, C. P. W., & Loh, P. H. (2004). Influence of dietary simulating solvents on the hardness of provisional restorative materials. *Dental Materials, 20*(4), 370–376.
19. Akova, T., Ozkomur, A., & Uysal, H. (2006). Effect of food-simulating liquids on the mechanical properties of provisional restorative materials. *Dental Materials, 22*(12), 1130–1134.
20. Korkmaz, T., Dogan, A., & Usanmaz, A. (2005). Dynamic mechanical analysis of provisional resin materials reinforced by metal oxides. *Bio-Medical Materials and Engineering, 15*(3), 179–188.
21. Saygili, G., Sahmali, S. M., & Demirel, F. (2003). The effect of placement of glass fibers and aramid fibers on the fracture resistance of provisional restorative materials. *Operative Dentistry, 28*(1), 80–85.
22. Kim, S. H., & Watts, D. C. (2004). Effect of glass-fiber reinforcement and water storage on fracture toughness (KIC) of polymer-based provisional crown and FPD materials. *The International Journal of Prosthodontics, 17*(3), 18–22.

23. Eisenburger, M., Riechers, J., Borchers, L., & Stiesch-Schtolz, M. (2008). Load-bearing capacity of direct four unit provisional composite bridges with fibre reinforcement. *Journal of Oral Rehabilitation, 35*(5), 375–381.
24. Garoushi, S. K., Vallittu, P. K., & Lassila, L. V. (2008). Short glass fiber-reinforced composite with a semi-interpenetrating polymer network matrix for temporary crowns and bridges. *The Journal of Contemporary Dental Practice, 9*(1), 14–21.
25. Merten, H. A., & Wiltfang, J. (1998). Technical improvements with wire ligatures-a clinical comparison between different paste/paste systems. *Die Quintessenz, 49*(4), 351–357.
26. Balkenhol, M., Köhler, H., Orbach, K., & Wöstmann, B. (2009). Fracture toughness of cross-linked and non-cross-linked temporary crown and fixed partial denture materials. *Dental Materials, 25*(7), 917–928.
27. Rosentritt, M., Lang, R., Behr, M., & Handel, G. *Fracture performance of provisional crown and bridge restoration materials.* Retrieved 30 March, 2021, from https://multimedia.3m.com/mws/media/587302O/protemp-4-protemp-crown-clinical-results.pdf
28. Fan, J., Tsui, C. P., Tang, C. J., & Chow, C. L. (2006). 3D finite element analysis of the damage effects on the dental composite subject to water sorption. *Acta Mechanica Solida Sinica, 19*(3), 212–222.
29. Bastioli, C., Romano, G., & Migliaresi, C. (1990). Water sorption and mechanical properties of dental composites. *Biomaterials, 11*(3), 219–223.
30. Carfagna, C., Guerra, G., & Nicolais, L. (1983). Effects of postcuring and water sorption on the mechanical properties of composite dental restorative materials. *Biomaterials, 4*(3), 228–229.
31. Rakhshan, V. (2015). Marginal integrity of provisional resin restoration materials: A review of the literature. *The Saudi Journal for Dental Research, 6*(1), 33–40.
32. Bindl, A., & Mörmann, W. H. (1999). Clinical evaluation of adhesively placed Cerec endo-crowns after 2 years—Preliminary results. *The Journal of Adhesive Dentistry, 1*(3), 255–265.
33. Biacchi, G. R., & Basting, R. T. (2012). Comparison of fracture strength of endocrowns and glass fiber post-retained conventional crowns. *Operative Dentistry, 37*(2), 130–136.
34. Rocca, G. T., Sedlakova, P., Saratti, C. M., Sedlacek, R., Gregor, L., Rizcalla, N., & Feilzer, A. J. (2016). Fatigue behavior of resin-modified monolithic CAD-CAM RNC crowns and endocrowns. *Dental Materials, 32*(12), e338–e350.
35. Güngör, M. B., Bal, B. T., Yilmaz, H., Aydin, C., & Nemli, S. K. (2017). Fracture strength of CAD/CAM fabricated lithium disilicate and resin nano ceramic restorations used for endodontically treated teeth. *Dental Materials, 36*(2), 135–141.
36. Bindl, A., Lüthy, H., & Mörnmann, W. H. (2006). Strength and fracture pattern of monolithic CAD/CAM-generated posterior crowns. *Dental Materials, 22*(1), 29–36.
37. El-Damanhoury, H. M., Haj-Ali, R. N., & Platt, J. A. (2014). Fracture resistance and microleakage of endocrowns utilizing three CAD-CAM blocks. *Operative Dentistry, 40*(2), 201–210.
38. Lise, D. P., Ende, A. V., De Munk, Y., Suzuki, T. Y. U., Viera, L. C. C., & Meerbeek, B. V. (2017). Biomechanical behavior of endodontically treated premolars using differrent preparation designs and CAD/CAM materials. *Journal of Dentistry, 59*, 54–61.
39. Gresnigt, M. M., Özcan, M., van den Houten, M. L. A., Schipper, L., & Cune, M. S. (2016). Fracture strength, failure type and Weibull characteristics of lithium disilicate and multiphase resin composite endocrowns under axial and lateral forces. *Dental Materials, 32*(5), 607–614.
40. Riera i Jorrin, M. (2014). *Study of the risk of fracture and debonding in a tooth restored by a post and a composite core and another by an endocrown methodology using FEA.* Universite Catholique de Louvain. Retrieved 30 March, 2021, from http://upcommons.upc.edu/handle/2099.1/24469
41. Dejak, B., & Młotkowski, A. (2013). 3D-Finite element analysis of molars restored with endocrowns and posts during masticatory simulation. *Dental Materials, 29*, 309–317.
42. Zhu, J., Rong, Q., Wang, X., & Gao, X. (2017). Influence of remaining tooth structure and restorative material type on stress distribution in endodontically treated maxillary premolars: A finite element analysis. *Journal of Prosthetic Dentistry, 117*(5), 646–655.

43. Hasan, I., Frentzen, M., Utz, T.-H., Hoyer, D., Langenbach, A., & Bourauel, C. (2012). Finite element analysis of adhesive endocrowns of molars at different height levels of buccally applied load. *Journal of Dental Biomechanics, 3*, 1–11.
44. Chen, B., Ma, Y., Wu, K., Chen. H., Li. L., Liu, J., & Chen, Z. (2015). Influence of various materials on biomechanical behavior of endocrown-restored, endodontically treated mandibular first molar: A 3D-finite element analysis. *Journal of WuHan University Technology. Materials Science Edition, 30*(3), 643–648.
45. Lin, C.-L., Chang, Y.-H., & Pai, C.-A. (2011). Evaluation of failure risks in ceramic restorations for endodontically treated premolar with MOD preparation. *Dental Materials, 27*, 431–438.
46. Lin, C.-L., Kuo, W.-C., Yu, J.-J., & Huang, S. F. (2013). Examination of ceramic restorative material interfacial debonding using acoustic emission and optical coherence tomography. *Dental Materials, 29*, 382–388.
47. Li, H., Li, Y., Yun, X., & Fok, A. S. L. (2010). Acoustic emission measurement of micro-debonding in composite restorations during polymerization. *Dental Materials, 26*(2), 131.
48. Li, H., Li, J., Yun, X., Liu, X., & Fok, A. S. L. (2011). Non-destructive examination of interfacial debonding using acoustic emission. *Dental Materials, 27*(10), 964–971.
49. Liu, X., Li, H., Li, J., Lu, R., & Fok, A. S. L. (2011). An acoustic emission study on interfacial debonding in composite restorations. *Dental Materials, 27*(9), 934–941.
50. Yang, B., Gou, J., Huang, Q., Heo, Y., Fok, A., & Wang, Y. (2016). Acoustic properties of interfacial debonding and their relationship with shrinkage stress in Class-I restorations. *Dental Materials, 32*, 742–748.
51. Lee, S.-Y., Lin, C.-T., Keh, E.-S., Pan, L.-C., Huang, H.-M., Shih, Y.-H., & Cheng, H.-C. (2000). Laser-induced acoustic emission in experimental dental composites. *Biomaterials, 21*(13), 1399–1408.
52. Vallittu, P. K. (2002). Use of woven glass fiber to reinforce a composite veneer. A fracture resistance and acoustic emission study. *Journal of Oral Rehabilitation, 29*(5), 423–429.
53. Choi, N.-S., Gu, J.-U., & Arakawa, K. (2011). Acoustic emission characterization of the marginal disintegration of dental composite restoration. *Composites Part A Applied Science Manufacturing, 42*(6), 604–611.
54. Kim, K.-H., Park, J.-H., Imai, Y., & Kishi, T. (1991). Fracture toughness and acoustic emission behavior of dental composite resins. *Engineering Fracture Mechanics, 40*(415), 811–819.
55. Mirsayar, M. M. (2018). On fracture analysis of dental restorative materials under combined tensile-shear loading. *Theoretical and Applied Fracture Mechanics, 93*, 170–176.
56. Yi, Y.-J., & Kelly, J. R. (2011). Failure responses of dental porcelain having three surface treatments under three stressing conditions. *Dental Materials, 27*, 1252–1258.
57. Ortega, V. I., Kaplan, A., Gomez, M. P., & Pumarega, M. I. L. (2015). Characterization of metal/ceramic interfaces in dental materials by acoustic emission. *Procedia Material Science, 8*, 683–692.
58. Ortega, V. I., Pumarega, M. I. L., Gomez, M. P., Kaplan, A., & Docimo, L. (2015). Adhesion study in metal-ceramic systems of dental restoration by acoustic emission. *Procedia Material Science, 9*, 477–483.
59. Skal's'kyi, V. R., Makeev, V. F., Stankevych, O. M., Kyrmanov, O. S., & Vynnyts'ka, S. I. (2015). Alternation of the types of fracture for dental polymers in different stages of crack propagation. *Materials Science*, 50(6), 836–843.
60. Skal's'kii, V. R., Makeev, V. F., Stankevich, O. M., Kyrmanov, O. S., Vynnyts'ka, S. I., & Opanasovich, V. K. (2015). Strength evaluation of stomatologic polymers by wavelet transform of acoustic emission signals. *Strength of Materials, 47*(4), 566–572.
61. GOST 25.506-85 *Design, calculation and strength testing. Methods of mechanical testing of metals. Determination of fracture toughness characteristics under the static loading*. Retrieved 10 January, 2021 from http://docs.cntd.ru/document/1200004652
62. Pestrikov, V. M., & Morozov, E. M. (2002). *Mechanika razrusheniya tverdykh tel (Fracture Mechanics of Solids)*. Professiya, St. Petersburg
63. Balkenhol, M., Ferger, P., Mautner, M. C., & Wöstmann, B. (2007). Provisional crown and fixed partial denture materials: Mechanical properties and degree of conversion. *Dental Materials, 23*, 1574–1583.

64. Kerby, E. R., Knobloch, L. A., Sharples, S., & Peregrina, A. (2013). Mechanical properties of urethane and bis-acryl interim resin materials. *Journal of Prosthetic Dentistry, 110*(1), 21–28.
65. Balkenhol, M., Meyer, M., Michel, K., Ferger, P., & Wöstmann, B. (2008). Effect of surface condition and storage time on the repairability of temporary crown and fixed partial denture materials. *Journal of Dentistry, 36*(11), 861–872.
66. *SKOP-8M portable 8-channel acoustic emission device for non-destructive testing of materials, products and structure elements*. Retrieved 25 March, 2021 from https://www.ipm.lviv.ua/our_developments/our_developments.php
67. Stankevych, O., & Skalsky, V. (2016). Investigation and identification of fracture types of structural materials by means of acoustic emission analysis. *Engineering Fracture Mechanics, 164*, 24–34.
68. Nazarchuk, Z., Skalskyi, V., & Serhiyenko, O. (2017). *Acoustic emission*. Springer International Publishing AG.
69. Stankevych, O., & Skalsky, V. (2017). The vibration of a half-space due to a buried mode I crack opening. *Wave Motion, 72*, 142–153.
70. Skalskyi, V., Makeev, V., Stankevych, O., & Pavlychko, R. (2018). Features of fracture of prosthetic tooth-endocrown constructions by means of acoustic emission analysis. *Dental Materials, 34*(3), e46–e55.
71. Ereifej, N., Silikas, N., & Watts, D. C. (2008). Initial versus final fracture of metal-free crowns, analyzed via acoustic emission. *Dental Materials, 24*, 1289–1295.
72. Liu, X., Fok, A., & Li, H. (2014). Influence of restorative material and proximal cavity design on the fracture resistance of MOD inlay restoration. *Dental Materials, 30*, 327–333.
73. Wang, Y., & Darvell, B. W. (2007). Failure mode of dental restorative materials under Hertzian indentation. *Dental Materials, 23*, 1236–1244.
74. Andreykiv, O. Y., Lysak, M. V., Serhiyenko, O. M., & Skalsky, V. R. (2001). Analysis of acoustic emission caused by internal crack. *Engineering Fracture Mechanics, 68*(11), 1317–1333.

Chapter 6
Rating of Hydrogen Damaging of Steels by the Wavelet Transform of Magnetoelastic Acoustic Emission Signals

To intensify the transportation of large volumes of products (oil, gas, etc.), it is necessary to ensure the reliability of main pipelines to prevent their failures and accidents [1–3]. Corrosion is the main cause of such failures during their operation (according to statistics, 80% of all failures) [4–7]. Dangerous factors of corrosion-mechanical fracture of construction include the aggressive role of hydrogen, which can cause uncontrolled failure of the construction in a short period of time [8–10]. The composition of the transporting medium (oil, gas) and the operation conditions of the pipelines contribute to the hydrogenation of the pipe material and, consequently, their fracture due to hydrogen brittleness.

In order to ensure the uninterrupted and safe supply of oil and gas products, it is necessary to maintain the pipeline systems in good working condition and to monitor their technical state. For this purpose, it is important to develop appropriate methods for monitoring and diagnosing the state of the pipe, separate components and units of technological equipment. The most common and well-developed among the modern NDT methods for the detection of operational defects of pipelines are acoustic methods, in particular the AE method [11]. Since for the traditional implementation of the AE method it is necessary to apply an external load to the object under control, which is not always acceptable in the conditions of pipelines operation, the MAE method is an alternative [12, 13]. It is possible to develop methods for efficient detection and quantitative evaluation of the degree of degradation of structural materials, particularly, under the influence of the hydrogen-mechanical factor [14].

In this chapter, the peculiarities of MAE signals by WT parameters during remagnetizing of structural steels with different degrees of hydrogenation are analyzed.

V. Skalskyi et al., *Acoustic Emission*, Foundations of Engineering Mechanics,
https://doi.org/10.1007/978-3-031-11291-1_6

6.1 Some Aspects of Operation of the Technical Systems in Hydrogenous Medium

6.1.1 Peculiarities of Oil and Gas Pipelines Operation

Such pipeline systems are subjected to the permanent influence of an aggressive environment—oil or gas. Let us consider some features of the operating conditions for each of them.

Carbon (79.5–87.5%) and hydrogen (11.0–14.5% of the oil weight) are main components of *oil*. The components of crude oil cause several active media inside the pipeline, in particular the surface-active ones. Such media are formed by surfactants in hydrocarbons or water, which interact with metal. Also, there are chemical and electrochemical active media, formed by melts of salts, aqueous solutions of electrolytes, which cause electrochemical or chemical corrosion [15]. According to the experts' data (although most of the causes of failures of oil refineries in Ukraine are due to human factors [16]), a significant percentage of accidents (23.5%) occur because of corrosive damage of pipes, shutoff and control valves [17, 18].

Methane is the base of the natural gas (98%). Gas, transported by main pipelines, is generally considered to be inert to corrosion environment, but in real operating conditions due to fluctuations in gas temperature and environment, gas condensation is possible on the inner surface, while salts, organic and sulfur-containing substances, CO_2 and bacteria present in it albeit in small quantities, can initiate and intensify the internal corrosion of the main gas pipelines.

The character and corrosion rate of main industrial pipelines are subjected to the influence of various factors. In particular, they are physical and chemical properties and composition of the products being transported; pH of the medium; the presence of organic and inorganic substances that can form protective films on the metal surface; oxygen content; chemical reagents used in technological processes of gas production and preparation; rate and mode of movement of the gas–liquid mixture, its pressure and temperature; abrasive particles in the liquid flow; vital activity of microorganisms that can cause biocorrosion of the pipe inner surface; chemical composition and structure of steel, its surface state; stresses in the metal, including residual; defects of metal and welded joints, etc.

Most often, the corrosive damages of the inner surfaces of industrial gas pipelines are localized in the lower part of the pipe because of the effect of the aggressive environment. Corrosion of the upper part is typical for gas pipelines of continental and marine deposits when conditions promote rapid condensation of wet gas in the colder part of the pipe. This leads to the formation of a water film in the upper part of the pipeline.

As a result of adsorption of some substances by the pipe surface, its strength and plasticity during long-term loading can change significantly. This ultimately leads to a decrease in the energy of initiation and propagation of surface cracks.

Welded joints are a complex inhomogeneous and thermodynamically unstable pipeline system. Due to differences in the chemical composition and structure of separate zones, the presence of residual stresses and plastic deformations in them there occurs an increased macro- and microelectrochemical inhomogeneity compared to the basic metal. Dynamic loads form conditions for fracture, which together with structural stresses cause the initiation of cracks. The localization of corrosion, most often caused by the microinhomogeneity of the metal, is one of the causes of the formation of microcavities or pittings, which grow into micro- and macrocracks.

Thus, during operation metal of main and industrial pipelines always corrode with the initiation and propagation of defects under the influence of external and internal factors and therefore are destroyed.

6.1.2 Specific Features of Operation of Feed Water Pipelines in High-Pressure Power Units

Most of the ultrahigh-pressure power (UHP) units of thermal power plants (TPP) of Ukraine have worked out the estimated lifetime and are operated in difficult conditions (unstable schedule of electrical loads, increasing duration of the unit downtime, use of non-design fuel, and of surface water sources polluted by industrial emissions, acute shortage of pipes, metal, ion exchange materials, etc.) [19]. To ensure the necessary reliability and to extend the life of the main thermal power equipment, it is necessary to consider a few problems. The most important of them are corrosive and corrosion-mechanical damages of the metal of various elements of the water–steam circuit of UHP units, caused by the operating environment.

Recently, such damages occur in the main feed water pipelines (FWP) units. The damage of the FWP metal is affected by both the quality of the coolant and compliance with the operating conditions of the used water-chemical regime. Damages have been recorded on both curved and straight sections of pipes.

All the UHP units use water regimes for oxygen dosing, namely neutral-oxygen and oxygen-ammonia. The effectiveness of these regimes is determined mainly by one condition—the high quality of the working environment. Disregard of this condition (deterioration of standard quality indicators of the environment, due to the presence of various non-standardized (especially organic) impurities) causes metal corrosion, and the need to replace the structural element may arise long before its lifetime is exhausted.

As a result of FWP long-term operation, there appears the accumulation of microdamages in the metal and, first of all, of its surface layers caused by the working environment. The influence of the stops and start-ups of units on these layers is associated with at least two factors: damage of oxide films on the steel surface; reduction of the quality of the working environment during the units start-up [20].

Various defects are found on the inner surfaces of the FWP of units: corrosive cavities, corrosion and corrosion-fatigue cracks, unbranched cracks with a narrow end, mainly oriented along the axis of the pipe [20, 21]. The character of the crack propagation indicates that the main damaging factors include the combined action of the working environment and fatigue, although their contribution to the fracture is different.

The results of investigations on the establishment of fracture mechanisms of FWP metal presented in the literature [20] indicate that high concentrations of carbon and other elements (S, Cl, O_2) are found in the cavities of cracks of the damaged metal from different thermal power plants, which, obviously, are part of the impurities of the coolant unit. These results show the possibility of oxidation of organic matter under conditions of neutral-oxygen or oxygen-ammonia water regimes, when there is a chemical bond between carbon and oxygen in one compound, present in the crack or corrosive cavity. The distribution of a number of elements in the crack cavity and on the adjacent metal surface has been also studied. It has been found that the maxima of the scans of the distribution of elements, carbon first of all, as well as oxygen, hydrogen, silicon, nitrogen, and chlorine, are grouped on the imaginary axis of the crack. This phenomenon indicates their directed diffusion from the environment (feed water) into the cavity of the growing corrosion crack.

6.1.3 Mechanism of Hydrogenation of the FWP Metal

Studies performed in [20] have shown an important role of organic impurities, available in the feed water in the damage of the FWP metal. This fact confirms the high carbon content in the crack cavities. In case there is no oxygen in the heat carrier, the role of organic matter in the UHP units is relatively passive. In the conditions of neutral-oxygen or oxygen-ammonia water regimes, we can see the increased corrosivity of organic impurities, which in the presence of oxygen oxidize, and thermally decompose with increase of the temperature of the working environment in the unit. Many products of both oxidation and thermal decompositions of organic matter are characterized by a high corrosivity. In the simplest case, the oxidation reaction of organic matter in an aqueous medium is accompanied by the formation of carbonic acid. Carbonic and other acids (acetic, formic, propionic, etc.) in the crack cavity (especially at its tip) contribute to the formation of an acidic environment, which inevitably activates corrosion processes, metal hydrogenation near the crack edges and the most—in the prefracture zone near its tip. Excessive hydrogenation and oxidation of FWP is an inevitable consequence of active corrosion reactions (especially in acidic environments), because of which hydrogen is formed.

The diffusion of hydrogen, as the lightest and the most mobile of all elements, into the metal is primary. It runs mainly along the grain boundaries, and local supersaturation of the metal with hydrogen weakens the intergranular and interatomic bonds, “separates” the boundaries and blocks of mosaics. As a result, further migration of oxygen and other impurities (C, Cl, N, S, P, Si, K, Ca) absorbed by the metal surface to the metal depth is possible. Due to exceeding a certain critical concentration, a new phase is released from the supersaturated solid solution in the form of CH_4, H_2S, H_2, oxides, and, less frequently, sulfides, nitrides, hydrides, and so on. Formation during operation of the second phase—a quaternary inclusion—significantly accelerates the metal fracture due to the wedging effect, with a simultaneous increase of the stress fields near the crack tips. A compression zone appears around the quaternary inclusion, which is surrounded by a tension zone. Such an interfacial boundary is usually incoherent, that is why a significant number of defects in the crystal lattice are concentrated here. Together with them, the substitution impurities (primarily hydrogen) are concentrated here. Therefore, even small stresses in these zones can cause delamination on the intergranular surface, i.e., contribute to the initiation of cracks. Due to the continuation of the metal work with quaternary inclusions, a significant increase in the crack growth rates should be expected, especially in the permanent presence of organic impurities in the feed water and continuous enrichment with them (products of their oxidation and thermolysis) of the near-surface layers of metal.

Considering the significant influence of metal hydrogenation on the rate of crack propagation, it is important to develop appropriate NDT methods for assessing the state and residual life of the product. Such measure allows to extend the time of safe operation of the pipelines and their elements.

6.2 Method for Estimating the Hydrogen Damage of Structural Materials by Wavelet Transform of Magnetoelastic AE Signals

6.2.1 Theoretical Bases of the MAE Signals Analysis by WT

As practice shows, although in most signals the most important information is contained in the low-frequency domain, the decomposition of the approximation coefficients can lead to the loss of information concentrated in the high-frequency domain. Although the most important information of the signals mainly belongs to the low-frequency domain, the decomposition process of the approximation component may lead to the loss of information located in the high-frequency domain. Therefore, the DWPT is used for the investigation of the MAE signals. In the DWPT, a wavelet detail component is also further decomposed to obtain its own approximation and detail components.

After studying the properties and characteristics of different wavelets and taking into account the literature data [22], a wavelet sym8, which is almost symmetrical and orthogonal, has been chosen (Fig. 6.1).

Since the discretization frequency of the MAE signals is 2 MHz, the maximum frequency present in the signal is 1 MHz. Three levels of signal decomposition are chosen (Fig. 6.2) that correspond to eight frequency ranges (Table 6.1). The frequency range is 0.2–0.6 MHz; i.e., wavelet packets marked $W1$–$W5$ are chosen.

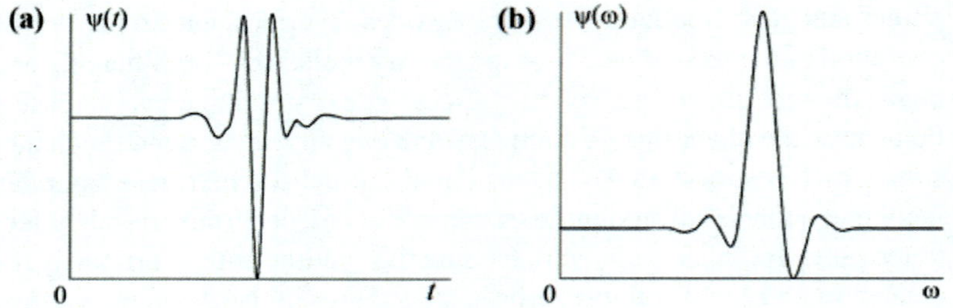

Fig. 6.1 Schematic representation of the wavelet function sym8 (**a**) and its Fourier transform (**b**)

Table 6.1 Correspondence of wavelet packets with frequency ranges

№.	Wavelet packet	Frequency range, kHz	Designations
0	(3.0)	0–125	*W*1
1	(3.1)	125–250	*W*2
2	(3.2)	250–375	*W*3
3	(3.3)	375–500	*W*4
4	(3.4)	500–625	*W*5
5	(3,5)	625–750	–
6	(3.6)	750–875	–
7	(3.7)	875–1000	–

The MAE signals are analyzed by the energy criterion, because differences in energy distribution between the components of the decomposition of WT can indicate a certain type or mechanism of restructuring the domain structure during the ferromagnet remagnetizing.

The total energy of the MAE signal is determined by formula (1.7), and its distribution by wavelet packets—using Eq. (1.8). Having the distribution of energy at each level, we can trace the change in energy for each stage of restructuring the domain structure.

To evaluate the MAE signal energy parameter in terms of its amplitude, the following dependence is used [23]

Fig. 6.2 Packet wavelet transform tree

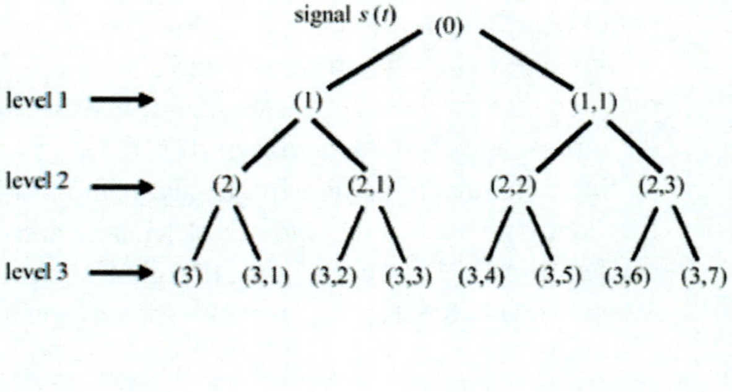

$$E_{\mathrm{MAE}} = \mathrm{const} \sum_i n_i A_i^2, \tag{6.1}$$

where n_i is a number of pulses with amplitude A_i.

The time-averaged local energy of the MAE signal pulses for a certain time interval $[t_1, t_2]$ by its CWT is determined by a known relation [24]:

$$E_{\mathrm{WT}}(a) = \|\mathrm{WT}_a(a, b)\|^2 = \frac{1}{N} \sum_{t_1}^{t_2} \mathrm{WT}^2(a, b), \tag{6.2}$$

where WT(a, b) is the value of the wavelet coefficient of the local pulse, N is the number of local averaging pulses.

The energy parameter of local pulse of the MAE signal is estimated using the previously proposed method (see Chap. 3), with application of Eq. (3.3). For the DWPT and for the determination of the energy parameters of the MAE signal, the toolbox written in Matlab7 (MathWorkInc.), developed by the authors, is used.

The first step is the determination of the dominant frequency ranges of MAE signals for the different magnetic field strength using DWPT. Next step is the determination of the energy and amplitude–frequency parameters of the MAE signal using CWT.

6.2.2 Materials for Experimental Research

To build a method for assessing the hydrogen damage of materials, such structural steels are chosen: hot-rolled carbon steel and high-quality low-carbon steel. Consider their characteristics.

The chemical composition (wt%) of *hot-rolled carbon steel* is given in Table 6.2. Strong and plastic ferrite is the basis of such steel structure (Fig. 6.3a).

Phosphorus and sulfur are referred to harmful impurities. Phosphorus forms a solution with ferrite, reducing the metal plasticity at high temperatures and increasing the brittleness at low temperatures.

For investigations, the prismatic samples of dimensions 240 × 30 × 2 mm are prepared. The microhardness of the sample material is determined by the device PMT-3M at a load of 0.49 N, according to the [25]. It is found that hot-rolled carbon steel has a low hardness (~ 140 HV). Macrohardness according to Rockwell is 70 HRB, according to Brinell—114 HB.

Table 6.2 Chemical composition of hot-rolled carbon steel

C	Si	Mn	S	P	Cr	Ni	Cu
0.3	0.35	0.65	0.04	0.03	0.28	0.25	0.45

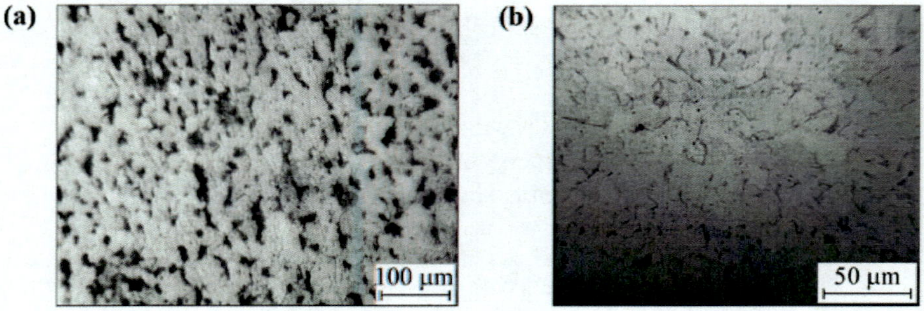

Fig. 6.3 Microstructure of hot-rolled carbon steel (**a**) and high-quality low-carbon steel (**b**)

Table 6.3 Chemical composition of the high-quality low-carbon steel

C	Si	Mn	S	P	Cr	Ni	Cu
0.15	0.17	0.65	0.04	0.035	0.25	0.25	0.25

High-quality low-carbon steel is used to manufacture parts that require high plasticity, surface hardness, and low core strength, the ability to operate at temperatures of − 40° to 450 °C. Its chemical composition (wt%) is given in Table 6.3.

Microstructure of the high-quality low-carbon steel (Fig. 6.3b) consists of deformed ferrite grains (size of 15–20 μm) and particles of perlite, which are placed along their boundaries, or in the form of individual inclusions (size of 5–10 μm).

Prismatic samples of dimensions 240×30×2 mm are prepared for investigations. The microhardness of steel is at the level of ~ 160 HV, and the macrohardness according to Rockwell is in the range of 71–72 HRB, according to Brinell—134 HB.

6.2.3 The Procedure of Experimental Research

The experiments are carried out in two stages. At the first stage, the specimens are magnetized in a solenoid (the number of loops of winding is 1500, the thickness of winding (5 layers of copper wire) is 35 mm, the length of winding wire is 94 mm) (Fig. 6.4).

The magnetizing frequency is 9 Hz. The MAE is recorded by a high-sensitive primary piezoelectric transducer with an operating frequency range of 0.2–0.6 MHz. MAE is investigated at different magnetic field induction B_{max} (T): 0.21; 0.45; 0.7; 0.9; 1.03.

After magnetizing, the specimens are hydrogenated in a special made chamber in a medium of hydrogen gas at high temperature. Figure 6.5 shows the modes of heating and hydrogenation of samples in the gas chamber. For receiving the optimum hydrogenation regime, the chamber with specimens is heated to 1138 K for time I; for

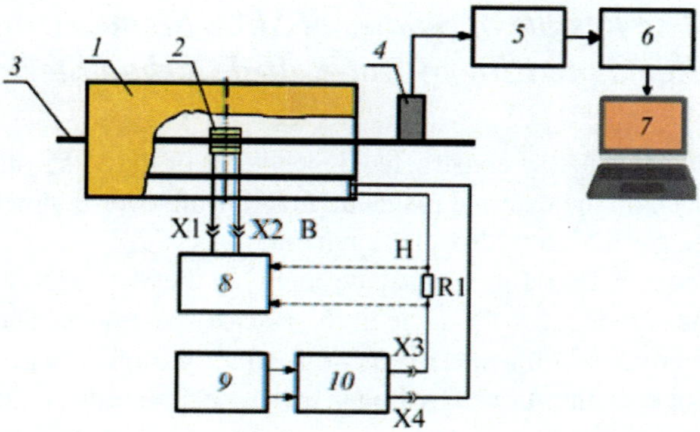

Fig. 6.4 Schematic diagram of the testing setup: *1*—solenoid; *2*—measuring winding; *3*—specimen; *4*—MAE measuring transducer; *5*—preamplifier; *6*—MAE device; *7*—personal computer; *8*—digital voltmeter; *9*—functional generator; *10*—low-frequency amplifier; *R*1—active load

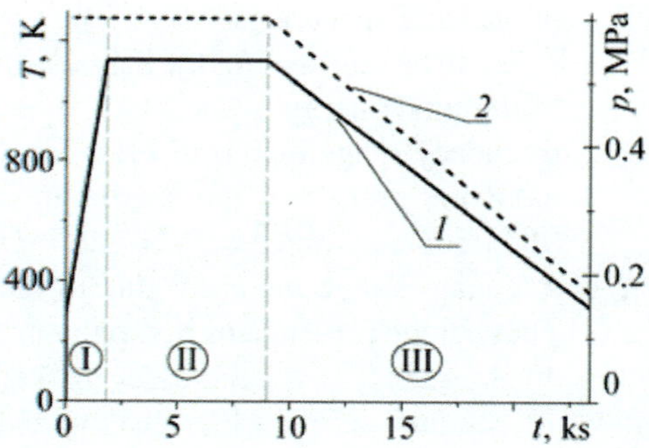

Fig. 6.5 Modes of heating and hydrogenation of samples: *1*—time dependence of temperature change; *2*—time dependence of pressure change

time II—the area is hydrogenated at constant temperature *T* and pressure *p*; III—the hydrogenated samples are cooled in the chamber together with the furnace.

Taking into account the literature data [12, 14], the hydrogenation modes are chosen so as to bring the concentration of hydrogen in the metal as close as possible to the real one in ferromagnetic structural materials (0.5–2.0 ppm) after their long-term operation. After hydrogenation, the test specimens are repeatedly tested in the solenoid. The dependences of the frequency and energy parameters of the MAE signals on the change of induction of the magnetic field and the hydrogenated state of specimens are compared.

In order to increase a precision of an estimation of the parameters of the MAE signals, we have used the multiple measurements for each value of magnetic field strength. The statistic processing of the recorded information [26] is contained in software of the measuring system.

6.2.4 The Results of Studies of MAE Signals During Remagnetizing of Hot-Rolled Carbon Steel Specimens

Figure 6.6 presents the constructed dependence of the MAE signal energy, obtained by Eq. (6.1), on the external magnetic field during over-magnetizing of as-cast steel specimens, 40 h after hydrogenation in air.

It can be seen from Fig. 6.6 that the energy of the MAE signals in the hydrogenated specimens is twice that the same in the as-cast specimens. The energy of the MAE signals decreases in the specimens 40 h after hydrogenation. It corresponds to the lowering of concentration of hydrogen in the specimens due to desorption of diffusion active hydrogen. However, the part of hydrogen stays in the microstructural traps of the material. Therefore, this energy does not reach the energy level of the MAE signals in the as-cast steel specimens.

Figure 6.7 shows the diagrams of the ratio of MAE energy by the various frequency ranges using DWPT by Eq. (1.8) for comparison of the different cases of magnetizing of the steel specimens.

It can be seen that for the magnetic field induction of 0.21 T the greatest percentage of the total energy of the MAE signals corresponds to the wavelet packet $W1$ (the frequency range of 0–125 kHz) for all types of specimens. For the magnetic field induction of 1.03 T, the maximum energy corresponds to the wavelet packet $W2$ (the frequency range of 125–250 kHz). The percentage of the MAE energy of this frequency range increases due to the presence of hydrogen in the steel structure. Thus, in the presence of hydrogen in the steel structure, the percentage of energy in this frequency range increases markedly.

The analysis of the local features of the MAE signals for the as-cast steel specimens and hydrogenated specimens is realized for the magnetic field induction of 1.03 T. For this purpose, we evaluate the signals in the time interval of 0–45 μs (Fig. 6.8a) and build CWT for them. Figure 6.8b shows the choice of the local data for the analysis (pulses *1*, *2*, *3*).

The maximum magnitude of the wavelet coefficient WT_{max}, the frequency f_{max} corresponding WT_{max}, the bandwidth Δf on the level $0.7WT_{max}$ at the WT–f projection (Fig. 6.9b) and the time t_{max} corresponding to the appearance of local pulse with

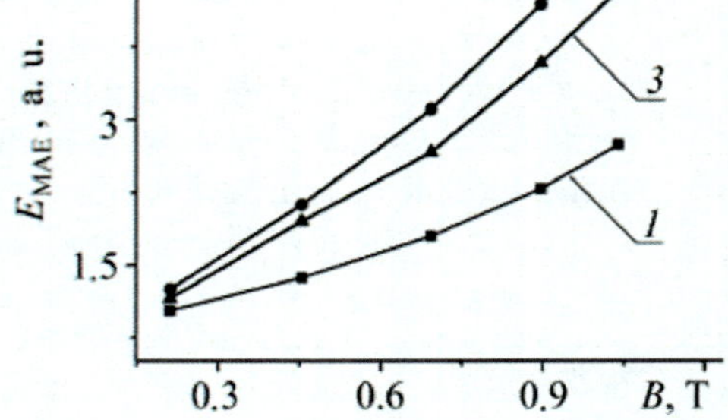

Fig. 6.6 Change in the total energy of MAE signals with increasing magnetic field induction for as-cast hot-rolled carbon steel specimens (*1*), after their hydrogenation (*2*) and 40 h after hydrogenation (*3*)

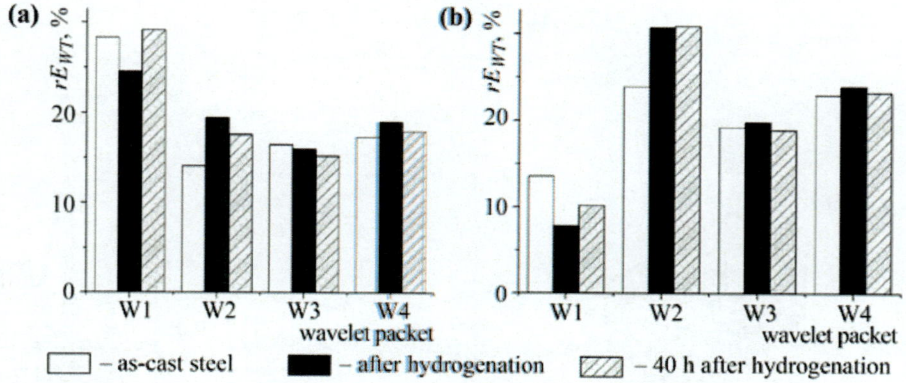

Fig. 6.7 Energy distribution of MAE signals by wavelet packets for as-cast hot-rolled carbon steel samples, after hydrogenation and 40 h after hydrogenation at different magnetic field induction: **a** $B = 0.21$ T; **b** 1.03 T.

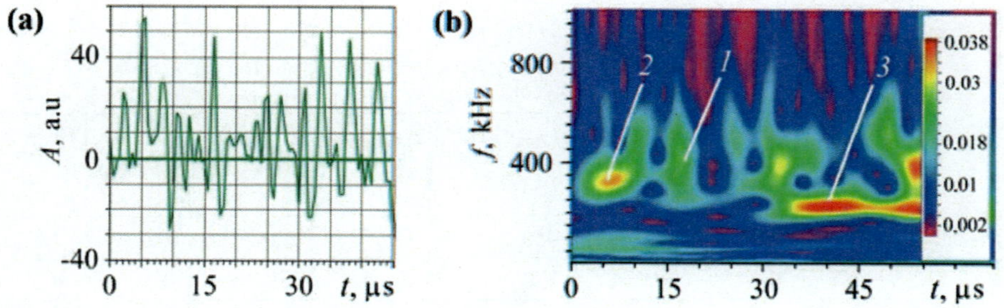

Fig. 6.8 Waveform of the typical MAE signal of the period 0–45 μs (**a**) and time–frequency dependence of CWT (**b**) for as-cast hot-rolled carbon steel at the magnetic field induction $B = 1.03$ T

WT_{max}, the time interval $[t_1, t_2]$ and the duration of radiation Δt at the level $0.5WT_{max}$ at the WT–*t* projection (Fig. 6.9c) are determined for each local maximum *1*, *2*, *3* (Fig. 6.8b) by the CWT.

The energy parameter for each local maximum E_{WT} is calculated by Eq. (3.3). Table 6.4 shows the result of calculations of the CWT parameters of the characteristic MAE signal for as-cast hot-rolled carbon steel. It is found (Table 6.4; Fig. 6.8b) that the local MAE pulses with different dominant frequencies alternate with each other, which, obviously, may indicate the alternation of the mechanisms of their generation.

A similar analysis of the local features of MAE signals is carried out for the hydrogenated specimens. The obtained data are averaged for each frequency range and dependences of the energy parameter E_{WT}, the radiation duration Δt and the frequency bandwidth Δf on the frequency f_{max} of the local maximum of MAE signal are plotted (Fig. 6.10).

It is found (Fig. 6.10) that as the dominant frequency decreases, the energy of local pulses, the duration of their radiation increase, and the bandwidth decreases. It is also concluded that in the presence of hydrogen in the metal to compare with the

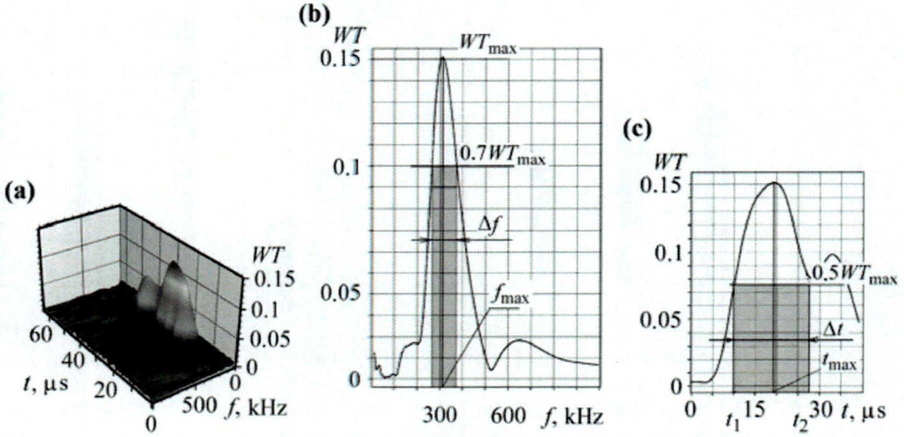

Fig. 6.9 Scheme for determining the CWT parameters for the local pulse **a** by projections on the planes WT–f, **b** (WT_{max}, f_{max}, Δf) and WT–t, **c** (t_{max}, $[t_1, t_2]$, Δt)

Table 6.4 Amplitude–frequency and energy parameters of local maxima of the MAE signal for the as-cast hot-rolled carbon steel (see Fig. 6.8b)

Number of local maximum	Wavelet packet	WT_{max}	f_{max}, kHZ	t_{max}, µs	Δf, kHz	Δt, µs	$E_{WT} \times 10^3$
1	*W*4	0.026	420	17.25	270	6	2.6
2	*W*3	0.034	330	6.25	130	10	8.0
3	*W*2	0.0385	230	40.25	65	16.5	18.0

as-cast material, the energy and duration of the local pulses of MAE signals increase (Fig. 6.10a, b), and the bandwidth decreases (Fig. 6.10c).

6.2.5 *The Results of Investigation of MAE Signals During Remagnetizing of High-Quality Low-Carbon Steel Specimens*

The peculiarities of the frequency characteristics of MAE signals are studied using DWPT. The total energy of MAE signals of steel specimens in the as-cast state and after hydrogenation for different magnetic field induction is evaluated by Eq. (6.1). It is found (Fig. 6.11) that with increasing the induction of magnetic field, the energy of MAE signals increases for both types of specimens. Similarly to high-quality low-carbon steel, hydrogen causes an increase in the energy of the MAE signals compared to the as-cast material. The obtained results agree with the dependences for the sum of the MAE signals amplitudes presented in [14].

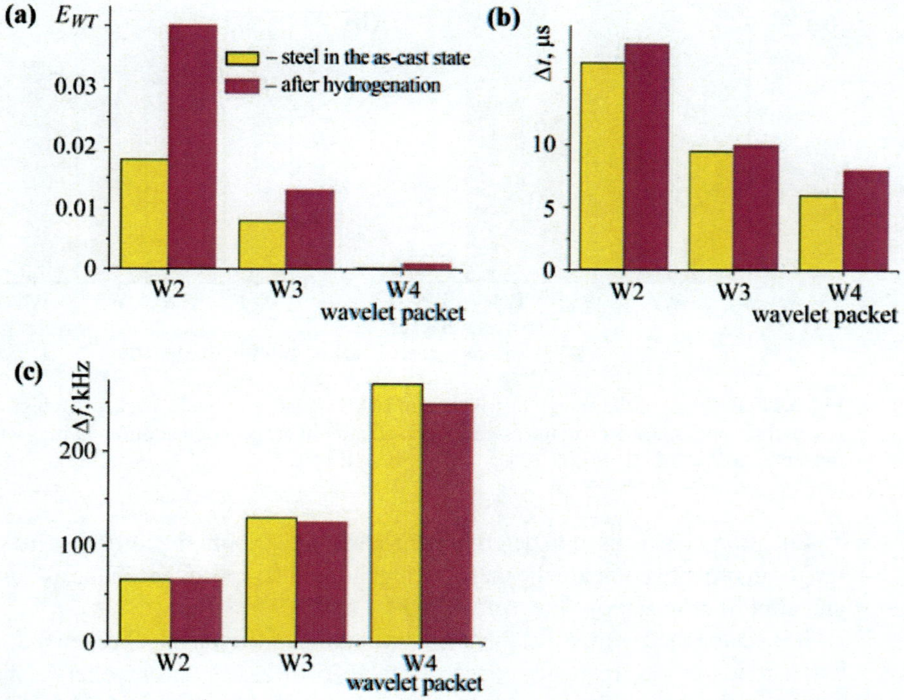

Fig. 6.10 Dependence of the energy parameter E_{WT} (**a**), the duration of radiation Δt (**b**), and the bandwidth Δf (**c**) of local events in the MAE signal on the frequency for high-quality low-carbon steel in the as-cast state and after hydrogenation

Fig. 6.11 Dependence of the energy of the MAE signals on the induction of magnetic field for high-quality low-carbon steel specimens: *1*—as-cast specimen; *2*—after their hydrogenation

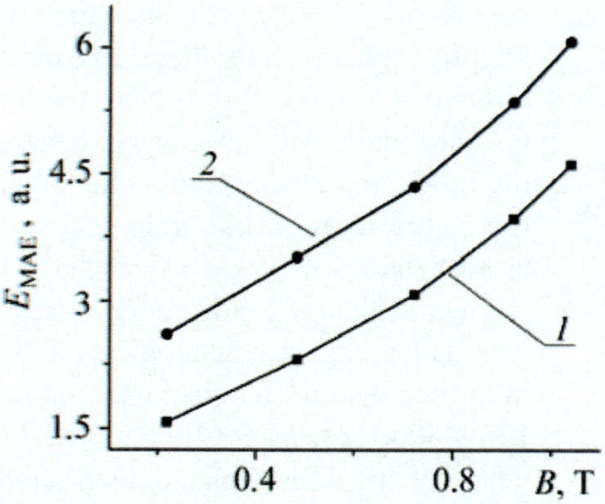

Thus, the increase in the energy of MAE signals during remagnetizing of structural materials compared to the standard indicates the presence of hydrogen in their structure. The energy of MAE signals can be used as an informative parameter when diagnosing the state of ferromagnetic materials.

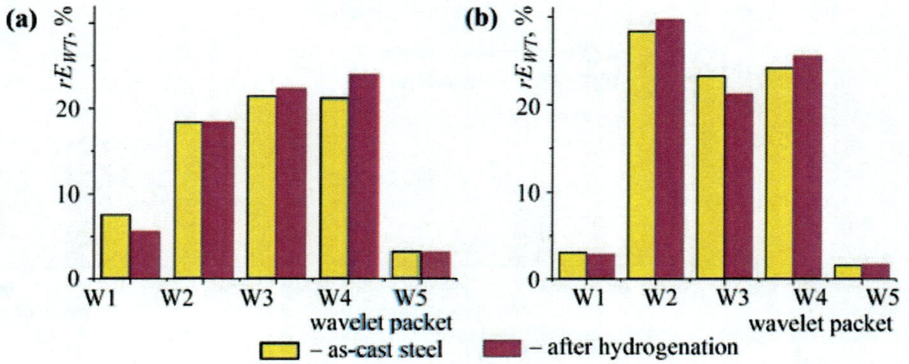

Fig. 6.12 Diagram of the energy distribution of MAE signals by wavelet packets for high-quality low-carbon steel specimens in the as-cast state and after hydrogenation at different magnetic field induction: **a** $B = 0.21$ T; **b** 1.03 T

Equation (1.8) is used to determine the energy distribution at different frequency ranges in order to compare the magnetizing of the steel specimens in the as-cast state and after hydrogenation (Fig. 6.12).

It is established (Fig. 6.12) that the contribution of the wavelet packet $W5$, which corresponds to the frequency range of 500–625 kHz, in the total energy of MAE signals is the lowest (less than 5%) with magnetic field induction 0.21 and 1.03 T. So, this frequency range is not considered in further studies. For the induction of a magnetic field of 0.21 T (Fig. 6.12a), the largest percentage of energy in the MAE signal for both as-cast steel and after hydrogenation is contributed by the wavelet packets $W3$ (the frequency range of 250–375 kHz) and $W4$ (the frequency range of 375–500 kHz). At the same time, in the hydrogenated material this percentage is slightly higher than for the as-cast material. As the magnetic field increases (magnetic induction 1.03 T) (Fig. 6.12b), the largest percentage of energy in the MAE signal is contributed by the $W2$ wavelet packet (the frequency range of 125–250 kHz), i.e., the lower frequencies than in the previous case dominate. In hydrogenated material, the contribution of this frequency range to the MAE signal is greater than for the as-cast steel.

Such a change in the energy distribution according to the frequency ranges with increasing magnetic field on the CWT of MAE signal is demonstrated by an increase in the concentration of local maxima at induction of 1.03 T in the frequency range of 125–250 kHz (Fig. 6.13c) compared to the CWT for induction 0.21 T (Fig. 6.13a), when the local maxima are distributed in a wide range of 250–500 kHz.

For the field induction of 1.03 T (contrary to 0.21 T induction), the values of wavelet coefficients corresponding to the local maxima increase significantly (see Fig. 6.13b, d).

To explain this effect based on the domain structure of ferromagnets, we present the following considerations. The equilibrium state of each domain wall determines its limiting value of the magnetic field induction at which it is in a steady state [27]. During the growth of the field, the formation of the MAE frequency spectrum

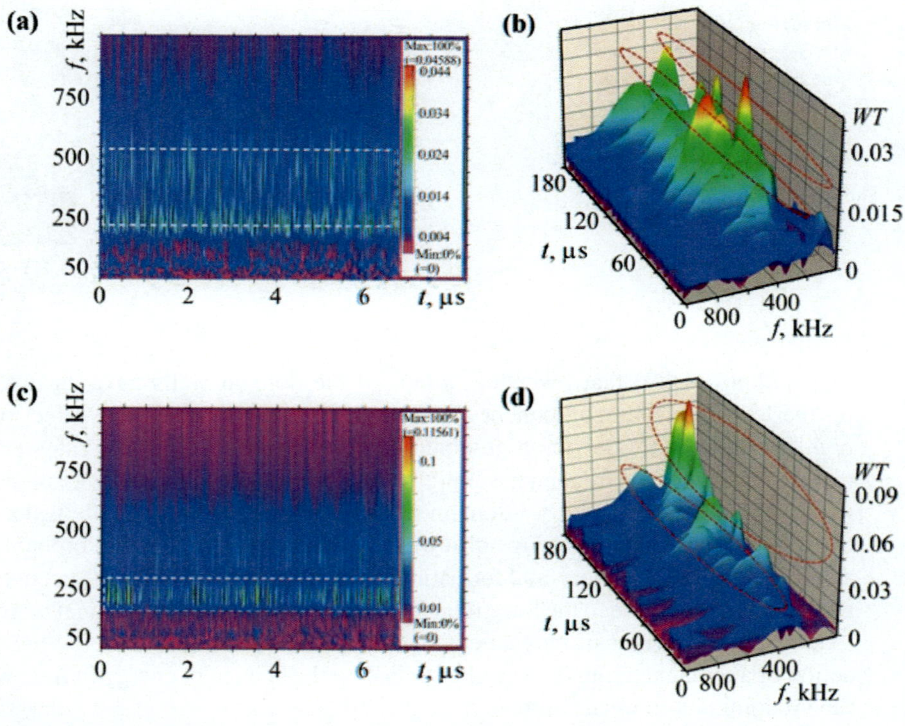

Fig. 6.13 CWT of typical MAE signals for samples from high-quality low-carbon steel at the magnetic field induction of 0.21 T (**a**, **b**) and 1.03 T (**c**, **d**): **a**, **c** the 2D-frequency projection of the period 0–8 μs; **b**, **d** the 3D-representation of the period 0–180 μs

involves first individual jumps of different size of the domain walls. Then there are processes of "cluster formation", the overlapping of jumps. As a result, the frequency of the spectrum decreases. Further, the spectrum is formed mainly by "clusters", which determines the absence of a further increase in the frequency of MAE signal generation. This fact can explain the shift of the maximum energy of the MAE signals to the lower frequency ranges for the high-quality low-carbon steel specimens in the as-cast state and after hydrogenation. Thus, the increase in the percentage of energy of the dominant frequency ranges in the presence of hydrogen in steel can serve as an informative parameter when diagnosing materials in contact with hydrogen.

The change in the energy parameter of signals within the predominant frequency range in time for the induction of a magnetic field of 1.03 T is also analyzed using Eq. (6.2) (Fig. 6.14). For the analysis, a time interval commensurable with the duration of Barkhausen noise is chosen [28], starting from the beginning of signal recording: 0–100; 100–200; and 200–300 μs. It is found (Fig. 6.14) that during remagnetizing of steel for short periods of time the dominant frequency ranges change: If at first the MAE signals have higher energy at the wavelet packets $W3$ (the frequency range of 250–375 kHz) and $W4$ (375–500 kHz), then with time the maximum energy is shifted toward the lower frequencies $W2$ (125–250 kHz).

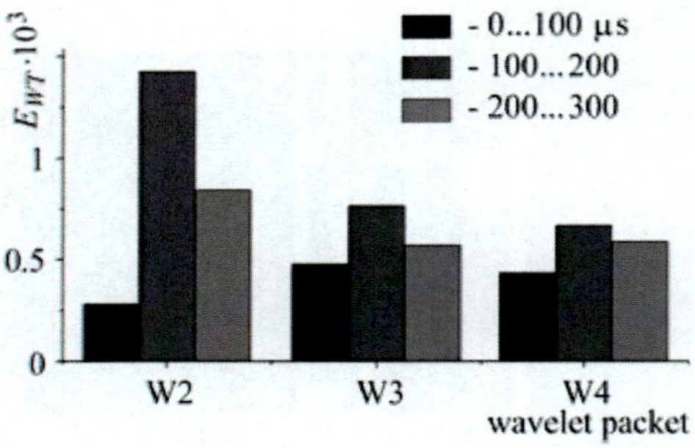

Fig. 6.14 Energy of the MAE signal for different time intervals within the dominant frequency range for the induction of a magnetic field of 1.03 T for as-cast high-quality low-carbon steel

It is known [29] that reversible jumps of the domain walls have an influence on the low-frequency component of MAE, while irreversible jumps affect on the high-frequency one. For the high-quality low-carbon steel, during remagnetizing, the range with higher frequencies initially has higher energy. It can be assumed that irreversible jumps of non-180° domain walls occur more intensively alongside with the reverse ones. Then the range of lower frequencies dominates. So, probably, the stage of "cluster formation" and formation/annihilation of domain walls occurs. As a result, the contribution to the low-frequency component of the MAE signal increases.

Thus, the dynamics of magnetic processes in a ferromagnetic material under the action of a remagnetizing field can be studied by the change in energy in time within the predominant frequency range.

According to the analysis of local pulses of MAE signals for high-quality low-carbon steel specimens after their hydrogenation, the same results are obtained as for hot-rolled carbon steel: Their energy parameters and duration increase, and the bandwidth decreases compared to the as-cast material. It is known that during propagation of the AE elastic waves, the narrowing of the frequency range and the shift toward lower frequencies of the dominant frequency occur due to an increase in the defect size [30, 31]. Thus, we can hypothetically assume that the narrowing of the MAE bandwidth indicates an increase in the number (size) of local sections of unidirectional domains, and hence the power of the jumps of their walls.

Thus, the parameters of local pulses of MAE signals—energy, radiation duration, bandwidth—can be classified as informative for assessing the damage of structural materials. The alternation of local pulses in different frequency ranges testifies to the change in the mechanisms (sources) of their generation.

6.3 Approbation of the Research Technique on Specimens of Long-Term Operated Pipe Steels

6.3.1 The Pipe of the Main Oil Pipeline

The material of the pipe of the main oil pipeline is investigated. The pipe (the diameter 720 mm, the thickness of the wall 10 mm) was in operation in 1971–2009 [32]. The chemical composition of the pipe material (wt%) is as follows: 0.18C–0.49Si–0.95Mn–0.035S–0.03P–0.11Cr–0.19Cu.

To prepare the specimens from the wall, the pipe is conventionally divided into three zones: the outer (*1*), which is about one-third of the thickness and covers the outer surface of the pipe, the middle—the next third in thickness (*2*) and inner (*3*), which covers the inner surface of the pipe. The rough specimens for the production of specimens are oriented along the axis of the pipe (Fig. 6.15a), the specimens are cut out from the bottom part (Fig. 6.15b), because it is the one most exposed to the effect of aggressive environment of the transportation products.

Prismatic specimens with a size of 240 × 30 × 2 mm are made from the rough specimens. During their manufacture, such treatment regimes are followed, in which minimal hardening and mechanical stresses are formed, and also there is a minimum change in the structure and phase state.

Note that the microstructure of the degraded metal consists of ferritic-pearlite columnar crystals [33]. According to the results of metallographic studies, it is found that the exploited metal is damaged along the cross section of the pipe wall by a significant number of pores with a size of 1–3 μm. Their number increases by 10–20% from the outer to the inner layer according to the pipe metal thickness. The pores formed during operation are of a correct spherical shape, which is typical for the defects caused by gases in the metal structure.

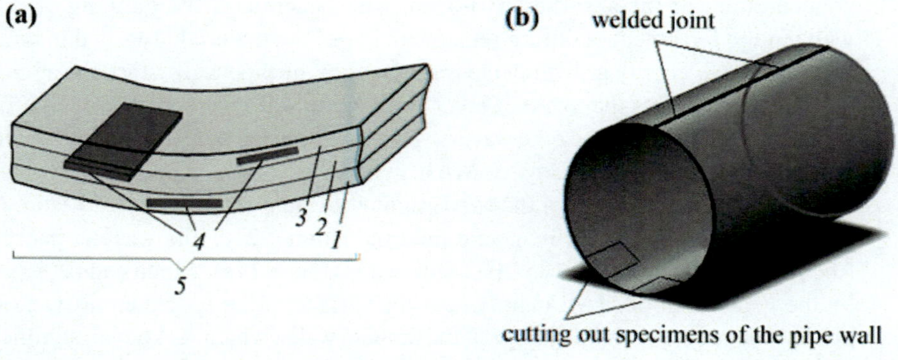

Fig. 6.15 Scheme of cutting out (**a**) of the rough specimens for making experimental specimens: *1*–*3*—the external, middle, and internal layer of the pipe wall, respectively; *4*—the places for cutting out specimens from the pipe wall; *5*—the pipe wall from the bottom part (**b**)

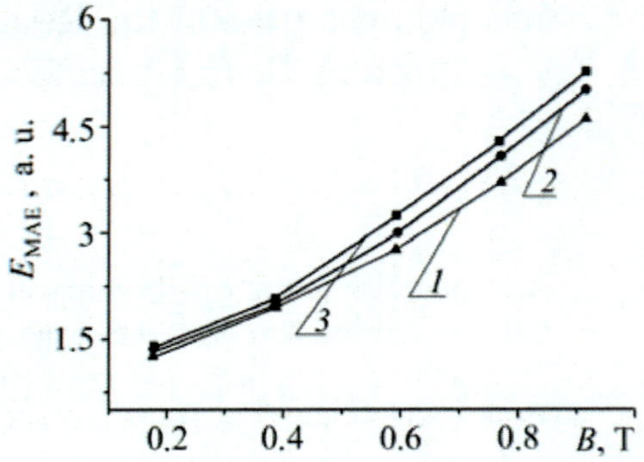

Fig. 6.16 Total energy of MAE signals versus the magnetic field induction for the specimens of the pipe material from the oil pipeline for various layers of the pipe wall: *1*—outer; *2*—middle; *3*—inner layer

Figure 6.16 presents the plotted dependence of the energy of MAE signals on the induction of the magnetic field during remagnetizing of samples of the oil pipeline cut out from the inner, middle, and outer layers of its wall.

It is found (Fig. 6.16) that MAE signals, generated during remagnetizing of specimens from the inner layer of the pipe wall have the greatest energy, and the lowest one has the specimens from the outer layer, which is consistent with the results of study [14, 33]. With an induction of a magnetic field of 1.03 T, the energy of MAE signals is 13% higher in the inner layer of the wall than in the outer one. This phenomenon can be explained by a number of factors. First, as revealed by microfractographic studies, the microstructure in different layers of the pipe wall is slightly different. In particular, a large number of pores are recorded in the inner part of the oil pipeline. It is known that the higher porosity of the material contributes to the reduction of MAE due to the demagnetizing effect of pores [34]. So, according to the analysis of literature sources in the inner pipe layer MAE should be the smallest. At the same time, the results of our experiments show the opposite effect. Comparing the obtained dependences of the energy of MAE signals with similar ones for the as-cast structural steel and after hydrogenation, we can assume that the presence of some concentration of the absorbed hydrogen, which facilitates the jumping of domain walls in the ferromagnet during remagnetizing of the material [14], is the dominant factor of MAE energy growth in the inner layer of the pipe wall. Since the aggressive medium less affects the external layer of the pipe wall, the energy of MAE in this layer is smaller than in the other layers. Thus, the damage of ferromagnetic materials can be estimated by the energy of MAE signals.

The frequency features of the MAE signals are analyzed by the WPT (Fig. 6.17). It can be seen that for the magnetic field induction 0.21 T the wavelet packet $W1$ (the frequency range of 0–125 kHz) dominates (Fig. 6.17a), which can be explained by the generation of MAE mainly near the surface of the specimen, as well as the predominance of reversible jumps of the domain walls, which, as known, significantly affect the low-frequency component of the MAE spectrum [29].

Due to increase of the magnetic field induction the energy of the wavelet packets $W2$ (the frequency range of 125–250 kHz), $W3$ (250–375 kHz), $W4$ (375–500 kHz) increase, and the wavelet packet has the maximum energy (Fig. 6.17b). This can be

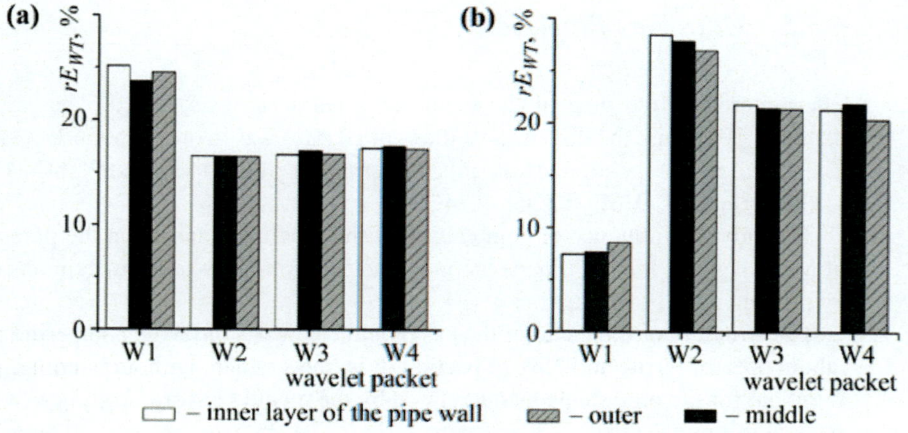

Fig. 6.17 Diagrams of the ratio of MAE energy by the WPT levels for the specimens of the various layers of the pipe wall of the oil pipeline for the various magnetic field induction: **a** $B = 0.21$ T; **b** 1.03 T

explained in the following manner. First, due to increase of the number of the irreversible jumps of the domain walls for the magnetic field induction 1.03 T the energy of the high-frequency component of the MAE signal increases, namely the percentage of the energy of the frequency range of 250–500 kHz increases. At the same time, the domain walls creation/annihilation process intensively develops. Thus, the energy of the low-frequency component of the MAE signal increases, that is the percentage of the energy of the frequency range of 125–250 kHz increases.

Also, for an induction of 1.03 T for specimens from the inner layer of the pipe, the percentage of energy of the dominant frequency range (of 125–250 kHz) is by 7% higher than from the outer one. This indicates a higher concentration of an adsorbed hydrogen, which is trapped in the microstructure of the degraded material.

Fig. 6.18 Total energy of MAE signals versus the magnetic field induction for the specimens of the pipe material of the gas pipeline for various layers of the pipe wall: *1*—outer; *2*—middle; *3*—inner layer

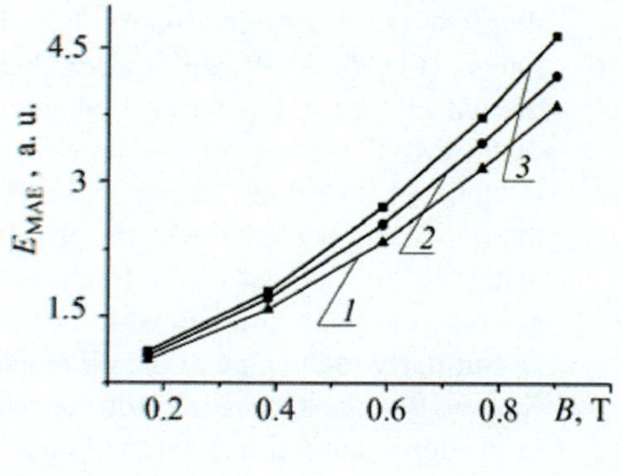

6.3.2 Main Gas Pipeline

The material of the pipe of the main gas pipeline is investigated. The pipe (the diameter 1000 mm, the thickness of the wall 10 mm) was in operation in 1973–2012. The chemical composition of the pipe material (wt%) is as follows: 0.16C–0.51Si–1.33Mn–0.04S–0.035P–0.32Cr–0.14Cu.

To study the influence of degradation of the pipeline material on the parameters of MAE signals, the specimens are manufactured similarly to the specimens of the oil pipeline (see Fig. 6.15).

The structure of the metal of the gas pipeline consists of ferrite and perlite grains (about 65% of ferrite and 35% of perlite). It is fine-grained, without banding, which is typical for the pipe steels manufactured by the method of forge-rolling. Averaged sizes of grains are ferrite—30–35 μm; perlite—20–25 μm.

The technological defects caused by the process of steel production (the nonmetallic phosphite inclusions formed during the final deoxidation of the liquid metal), the shrinkage cavities caused by shrinkage of the metal; a small amount of gas pores caused by the gas evolution during cooling of the metal to the solid phase, are found.

Figure 6.18 shows the dependences of the energy of the MAE signals on the magnetic field induction during remagnetizing of the specimens of the pipe of the gas pipeline cut out from the various layers of the pipe wall.

It can be seen that the MAE signals generated during remagnetizing of the specimens from the internal layer of the wall pipe have the greatest energy, and from the external layer of the wall pipe—the lowest. With the induction of a magnetic field of 1.03 T, the energy of MAE signals in the inner layer of the pipe wall is by 21% higher than in the outer. The reason for this effect can be a significant degradation of the metal of the internal layer of the pipe wall due to durable contact with the aggressive gas environment, as well as the presence of hydrogen in the metal structure adsorbed from the gas mixture.

To study the frequency features of the MAE signals, DWPT is used and the diagrams of energy distribution have been constructed according to the frequency ranges (Fig. 6.19). It can be seen that for the magnetic field induction of 0.21 T, the wavelet packet $W1$ (the frequency range 0–125 kHz) has the greatest energy of the MAE signals for all three specimens (Fig. 6.19a). For the magnetic field induction 1.03 T, the wavelet packet $W2$ (the frequency range of 250–500 kHz) has the maximum percentage of the energy of the MAE signals (Fig. 6.19b). Thus, the results of the analysis of the frequency features of the MAE signals for the oil and gas pipe material correlate with each other.

Similarly to the pipe of the oil pipeline for the induction of magnetic field 1.03 T (Fig. 6.19b), the hydrogen in the material structure increases the percentage of energy of the dominant frequency range for MAE signals of the inner layer of the pipe wall by 8.5% to compare with the outer one.

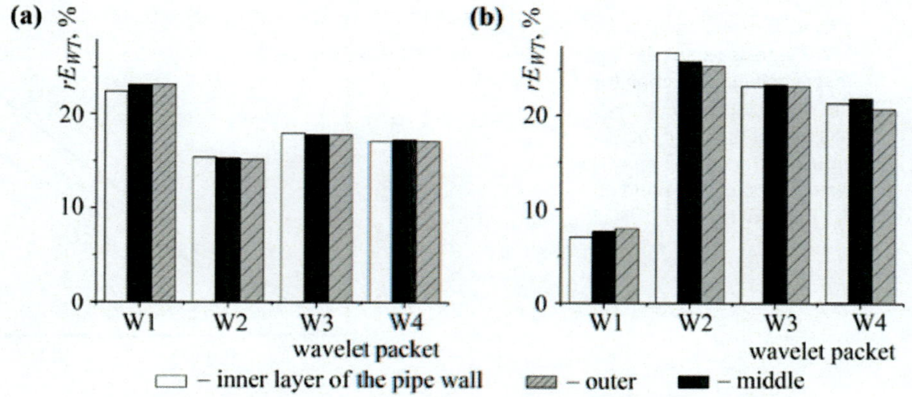

Fig. 6.19 Energy distribution of MAE signals by wavelet packets for samples from different layers of the gas pipeline wall at different magnetic field induction: **a** $B = 0.21$ T; **b** 1.03 T

6.3.3 *A Pipe for Feed Water of UHP Power Unit*

The material of the FWP, which is approaching the end of the project lifetime, is studied. In particular, the duration of the pipe operation is 120–150 thousand hours; number of starts—170–350; number of load cycles—850–1750. As a result of long-term operation, irreversible changes in the structure of the pipeline occur under the influence of mechanical, thermal, corrosive, and other factors that contribute to the initiation and propagation of cracks. Chemical composition (wt%) of the material is as follows: 0.13C–0.76Si–1.21Mn–0.04S–0.040P–0.16Cr–0.21Cu.

For researches, the rough specimens are prepared according to the scheme presented in Fig. 6.15 from a fragment of a pipe with a wall thickness of 46 mm, length of a segment arc of 280 mm, length of 480 mm.

As it is noted above, as a result of the action of corrosion-mechanical and other factors in the real FWP of UHP power units there appear the regions with varying degrees of metal damage. The spatial orientation and dimensions of corrosion cracks, as well as hydrogenated regions of metal, are different [35, 36]. Visual inspections of the material of pipe fragments, as well as cuts in the radial direction on the inner surfaces of the pipe wall reveal large and small macrocracks, layers of corrosion products, and so on. The cracks range in size from millimeters to several centimeters. They are oriented along the axis of the pipe. Depth of cracks in the radial direction is 0.1–5.0 mm.

According to the results of metallographic studies of materials of different layers of the pipe wall, it is established that the steel consists of ferritic grains and pearlite located along their boundaries. Average grain sizes are: ferritic—of 35–45 μm; pearlitic—of 15–20 μm. The material of the middle layer of the pipe wall has a slightly finer structure than the other two. It should also be noted that pearlitic inclusions do not have clear boundaries. This may be the result of decarburization under

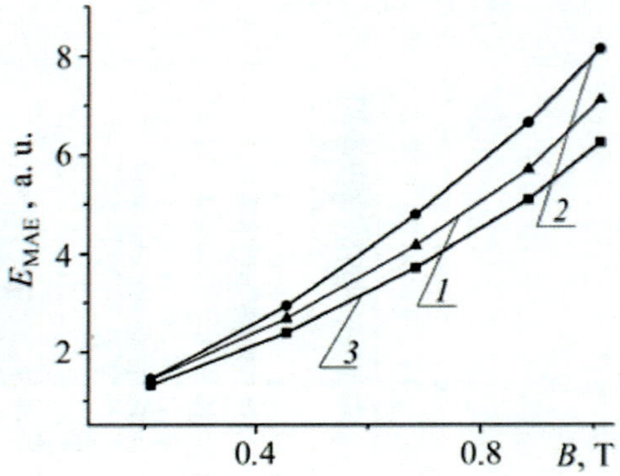

Fig. 6.20 Total energy of MAE signals versus the magnetic field induction for the specimens of the pipe material of the FWP for various layers of the pipe wall: *1*—outer; *2*—middle; *3*—inner layer

the influence of the working environment and elevated operating temperatures. The pores in the FWP material of the UHP power unit are not visualized.

The highest value of microhardness is recorded in the middle layer of the wall (185 HV). In the outer and inner layers, these values are—170 and 175 HV, respectively. The macrohardness of the metal of different layers of the wall of the FWP of the UHP power unit is the same according to both Rockwell and Brinell—82 HRB and 156 HB, respectively.

Figure 6.20 presents the dependence of the energy of MAE signals on the induction of the magnetic field during the remagnetizing of the samples of the FWP cut out from the inner, middle, and outer layers of its wall.

As we can see, the MAE signals generated during the demagnetizing of the specimens from the middle layer of the wall have the highest energy, and the lowest one—from the inner layer. For the induction of a magnetic field of 1.03 T, the energy of MAE signals in the middle layer of the wall is by 25% higher than in the inner. According to the mechanism of hydrogenation of the inner wall of the FWP of the UHP power unit, probably during the long-term operation of such a pipeline the hydrogen has penetrated deep into the pipe material, which intensifies corrosion processes and leads to a significant number of crystal lattice defects just in the middle layer of the wall. This fact is confirmed by the results obtained in [21], where the degree of volumetric damage of similar specimens is investigated by AE signals. The middle layer of the pipe wall is found to have the greatest degree of volumetric damage, followed by the outer and inner layers.

Therefore, summarizing the above-mentioned, we can assume that the highest energy of MAE signals for the middle layer of the pipe is caused by the highest concentration of hydrogen in it, which densely fills all the defects in the microstructure of the material and facilitates jumping of domain walls during remagnetizing by the external magnetic field.

According to the DWPT of MAE signals, the diagrams of energy distribution by frequency ranges are built and their frequency features are studied (Fig. 6.21).

With an induction of a magnetic field of 0.21 T, all the frequency ranges are present in the MAE signal of all three samples (Fig. 6.21a). The wavelet packet $W2$ (the frequency range of 125–250 kHz) has a slightly larger percentage of energy for

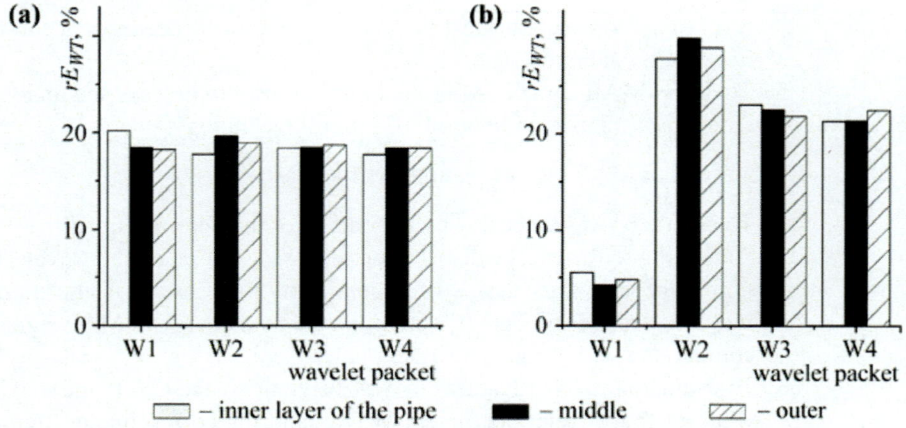

Fig. 6.21 Distribution of the MAE signals energy according to the WPT levels for the specimens of the various layers of the pipe wall of the FWP of UHP power unit for the various magnetic field induction: **a** $B = 0.21$ T; **b** 1.03 T

the MAE signals of the middle layer, and $W1$ (0–125 kHz)—for the inner one. As the magnetic field induction increases (1.03 T), the $W2$ wavelet packet (125–250 kHz) (Fig. 6.21b) has the maximum percentage of the MAE signal energy for all the types of specimens. The energy of the dominant frequency range is by 10% higher for the middle layer of the pipe wall than for the inner one.

According to the results of studies of the structural steels (hot-rolled carbon steel and high-quality low-carbon steel), hydrogen in the metal leads to an increase in the percentage of energy of the dominant frequency range. Regarding the energy distribution of the MAE signals of the middle layer, the wavelet packet $W2$ dominates (125–250 kHz) (Fig. 6.21), the percentage of which is larger to compare with the inner layer. Therefore, it can be argued that the middle layer of the pipe wall after long-term operation contains a higher concentration of residual hydrogen.

According to the results of complex studies of structural ferromagnetic materials, *the algorithm of assessing the state of their hydrogen damage by the parameters of MAE signals* is as follows:

1. To register MAE at the site of the object of control:
   - 1.1 To install primary transducers (PT) at the critical sites of the object (welds, high stresses zones, zones that have already been repaired, etc.).
   - 1.2 To select the maximum distance between the PT and the C-core magnet to measure the attenuation coefficient of acoustic waves.
   - 1.3 To check the serviceability of the equipment and calibrate its sensitivity.
   - 1.4 To analyze the background noises. All noises must be identified and minimized, and their parameters recorded. After adjusting the equipment and before working tests, to check for 15 min the noise background, which must be lower than the set threshold value. During the noise registration, the level of which exceeds the threshold, the noise source must be eliminated or the

level of discrimination must be increased and the procedures according to item 1.4 must be repeated.

1.5 To detect MAE signals. After the object control to process and analyze the data in accordance with the MAE-control technology.

2. The analysis of MAE signals parameters includes:

   2.1 To assess the MAE activity by the energy of the detected signals.
   2.2 According to the energy distribution of DWPT (based on the wavelet function sym8 and three levels of decomposition) to establish the dominant frequency ranges of signals; determine the total energy of MAE signals for comparison with reference values.
   2.3 To evaluate the local features of MAE signals by the CWT (the maximum value of the wavelet coefficient WT_{max} and the corresponding frequency f_{max}, the duration of radiation Δt and bandwidth of local maxima Δf, the energy parameter E_{WT}).

3. To classify the zones of the control object by the MAE activity using the signals' parameters.
4. To compare the obtained data with the reference data, registered in the laboratory.

Depending on the character of the change in the main parameters of MAE signals to assess the state of damage of the structural material and to make a decision on the necessity of additional control of damaged areas using other NDT methods.

References

1. Taylor, J. R. (2003). *Risk analysis for process plant, pipelines and transport*. Routledge.
2. Revie, R. W. (Ed.). (2015). *Oil and gas pipelines: integrity and safety handbook*. John Wiley & Sons.
3. Kiefner, J., & Trench, C. (2021). *Oil pipeline characteristics and risk factors: Illustrations from the decade of construction*. American Institute of Petroleum. Retrieved 30 March, 2021, from https://www.academia.edu/36648113
4. Papavinasam, S. (2013). *Corrosion control in the oil and gas industry*. Gulf Professional Publishing.
5. Fessler, R. R. (2008). Pipeline corrosion. Final report. Retrieved 30 March, 2021, from https://www.phmsa.dot.gov/sites/phmsa.dot.gov/files/docs/technical-resources/pipeline/gas-transmission-integrity-management/65341/finalreportpipelinecorrosion.pdf
6. Ossai, C. I., Boswell, B., & Davies, I. J. (2015). Pipeline failures in corrosive environments—A conceptual analysis of trends and effects. *Engineering Failure Analysis, 53*, 36–58.
7. Popoola, L. T., Grema, A. S., Latinwo, G. K., Gutti, B., & Balogun, A. S. (2013). Corrosion problems during oil and gas production and its mitigation. *International Journal of Industrial Chemistry, 4*, 35(2013). https://doi.org/10.1186/2228-5547-4-35
8. Nykyforchyn, H., Lunarska, E., Tsyrulnyk, O., Nikiforov, K., & Gabetta, G. (2009). Effect of the long-term service of the gas pipeline on the properties of the ferrite-pearlite steel. *Materials and Corrosion, 60*(9), 716–725.
9. Ohaeri, E., Eduok, U., & Szpunar, J. A. (2018). Hydrogen related degradation in pipeline steel: A review. *International Journal of Hydrogen Energy, 43*(31), 14584–14617.

10. Tsyrul'nyk, O. T., Voloshyn, V. A., Petryna, D. Yu., Hredil', M. I., & Zvirko, O. I. (2011). Degradation of properties of the metal of welded joints in operating gas mains. *Materials Science, 46*(5), 628–632.
11. Nazarchuk, Z., Skalskyi, V., & Serhiyenko, O. (2017). *Acoustic emission. Methodology and Application.* Springer International Publishing AG.
12. Shibata, M., & Ono, K. (1981). Magnetomechanical acoustic emission—A new method of nondestructive stress measurement. *NDT International, 14*(5), 227–234.
13. Jiles, D. C. (1988). Review of magnetic methods for nondestructive evaluation. *NDT International, 21*(5), 311–319.
14. Nazarchuk, Z. T., Andreikiv, O. Ye., & Skalskyi, V. R. (2013). *Otsinyuvannya vodnevoyi dehradatsii feromahnetykiv u mahnetnomu poli (Estimation of hydrogen degradation of ferromagnets in a magnetic field).* Naukova dumka.
15. Maruschak, P., Panin, S., Chausov, M., Bischak, R. T., & Chausov, N. G. (2018). Effect of long-term operation on steels of main gas pipeline: Structural and mechanical degradation. *Journal of King Saud University Engineering Science, 30*(4), 363–367.
16. Andrusyak, A. V. (2008). Analiz i prychyny vidmov na naftoprovodi ta faktory, yaki vplyvayut na ich ekspluataciyu (Analysis and causes of pipeline failures and factors affecting their operation). *Rozvidka ta rozrobka naftovych i hazovych rodovyshch (Exploration and Development of Oil and Gas Fields), 4*, 83–85.
17. Dmytrach, I. M., & Panasyuk, V. V. (1999). *Vplyv koroziinych seredovyshch na lokalne ruinuvannya metaliv bilya koncentratoriv napruzhen (Influence of corrosive media on local fracture of metals near stress concentrators).* PhMI.
18. Bolzon, G., Gabetta, G., & Nykyforchyn, H. (Eds.). (2021). *Degradation assessment and failure prevention of pipeline systems. Lecture notes in civil engineering* (Vol. 102). Springer.
19. Hyrshfeld, V. Y. (1980). *Rezhymy raboty i ekspluataciya teplovych elektrostanciyi (Operating modes and operation of thermal power plants).* Energiya.
20. Panasyuk, V. V. (Ed.). (2005). *Mechanika ruyinuvannya i micnist materialiv (Mechanics of fracture and strength of materials)* (Vol. 7). PH Akademperiodyka.
21. Skalskyi, V. R., & Andreikiv, O. Y. (2006). *Otsinka obyemnoyi poshkodzhenosti materialiv metodom akustychnoyi emisiyi (Estimation of materials volume damage by the acoustic emission method).* Publishing Center LNU im. I. Franka.
22. Lazorenko, O. V., Lazorenko, S. V., & Chernogor, L. F. (2007). Veyivlet-analiz modelnych signalov s osobennostyami. 2. Analiticheskoe i diskretnoye veyivlet-preobrazovaniya (Wavelet analysis of the model signals with peculiarities. 2. Analytical and discrete wavelet transforms). *Radio Physics and Radio Astronomy, 12*(3), 278–294.
23. Mirabile, M. (1975). Acoustic emission energy and mechanisms of plastic deformation and fracture. *NDT, 8*(2), 77–85.
24. Torrence, C., & Compo, G. P. (1998). A practical guide to wavelet analysis. *Bulletin of the American Meteorological Society, 79*(1), 61–78.
25. ASTM E384-17 (2017). *Standard test method for microindentation hardness of materials.* ASTM International.
26. Peters, C. A. (2001). *Statistic for analysis of experimental date.* AEESP.
27. Ng, D. H. L., Lo, C. C., & Jakubovics, J. P. (1994). The effect of demagnetizing and stray fields on magnetoacoustic emission. *Journal of Applied Physics, 75*(10), 7009–7011.
28. Spivak, L. V., Myasnikov, N. N. (2012). Veyivlet-analiz spektrov Barkkhausena v amorfnom splave $Fe_{78}B_{12}Si_9Ni_9$ (Wavelet analysis of Barkhausen spectra in an amorphous alloy $Fe_{78}B_{12}Si_9Ni_9$). *Vestnik Permskoho universiteta. Seriya: Fizika (Perm University Bulletin. Series: Physics), 1*(19), 61–64.
29. Kostin, V. N., Gurev, M. A., Vasilenkom, O. N., et al. (2013). Amplitudno-chastotnyye charakteristiki magnitoakusticheskoi emissii termoobrabotannych solavov zheleza (Amplitude-frequency characteristics of magnetoacoustic emission of heat-treated iron alloys). *Fizicheskaya mezomechanika (Physical Mesomechanics), 16*, 103–110.
30. Stankevych, O., & Skalsky, V. (2017). The vibration of a half-space due to a buried mode I crack opening. *Wave Motion, 72*, 142–153.

31. Andreykiv, O. Y., Lysak, M. V., Serhiyenko, O. M., & Skalsky, V. R. (2001). Analysis of acoustic emission caused by internal crack. *Engineering Fracture Mechanics, 68*(11), 1317–1333.
32. Skalskyi, V., Stankevych, O., & Dubytskyi, O. (2018). Estimation of effect of hydrogen on the parameters of magnetoacoustic emission signals. *Journal of Magnetism and Magnetic Materials, 454*, 375–385.
33. Skal's'kyi, V. R., Pochaps'kyi, E. P., Klym, B. P., Rudak, M. O., & Velykyi, P. P. (2016). Application of the method of magnetoelastic acoustic emission for the analysis of the technical state of 19G steel after long-term operation in an oil pipeline. *Materials Science, 52*, 385–389.
34. Gorkunov, E. S., Zakharov, V. A., Chulkina, A. A., & Ul'yanov, A. I. (2004). Internal demagnetization factor for porous ferromagnets in remagnetization process. *Russian Journal of Nondestructive Testing, 40*(1), 1–7.
35. Coulter, J. E., Stevens, D. M., Gehl, S., & Scheibel, J. (1988). Acoustic emission monitoring of fossil-fuel power plants. *Materials Evaluation, 46*(2), 230–237.
36. Vainman, A. B., Melechov, R. K., & Smiyan, O. D. (1990). *Vodorodnoye okhrupchivaniye elementov kotlov vysokogo davleniya (Hydrogen embrittlement of high pressure boiler elements)*. Nauk. dumka.